Biological Control

Intext Series in ECOLOGY

ARTHUR S. BOUGHEY, *Editor*
University of California, Irvine

Fundamental Ecology
Arthur S. Boughey

World Food Resources
Georg Borgstrom

Biological Control
Robert van den Bosch and P. S. Messenger

Water Resources
Nathaniel Wollman

Conservation
Archie S. Mossman

Biological Control

Robert van den Bosch
P. S. Messenger
University of California, Berkeley

Intext Educational Publishers
New York and London

Library of Congress Cataloging in Publication Data

Van den Bosch, Robert.
Biological control.

(Intext series in ecology)
Includes bibliographies.
1. Insect control—Biological control.
I. Messenger, P. S., joint author. II. Title.
SB932.V36   632′.7   73-1773
ISBN 0-7002-2433-5
ISBN 0-7002-2441-6 (pbk)

Intext Educational Publishers
257 Park Avenue South
New York, New York 10010

Contents

Series Preface

As we move further into the 1970's we are confronted with dire threats of imminent environmental disaster. While prophecies as to the actual doomsday vary from five years to thirty years from now, no professional ecologist seems willing to state categorically that mankind will survive into the next millenium unchallenged by any ecocatastrophe. Some indeed believe that before this time we and most of our familiar ecosystems are inevitably doomed to extinction.

Enough has now been said and written about such predicted disasters to instill in students, governments, and the public at large an uneasy feeling that something may be amiss. Terms such as *pollution, natural increase,* and *re-cycling* have begun to assume a realistic and more personal note as the air over our cities darkens, our rivers are turned into lifeless fire hazards, our domestic water becomes undrinkable, and we have to stand in line for any form of service or amenity.

Politicians, scientists, and the public have responded variously to this new situation. Tokenism is rampant in thought, word, and deed. Well-intentioned eco-activist groups have mushroomed, not only among youth, who are the most threatened as well as the most understanding segment of our societies. More specifically, in the restricted field of college texts, appropriate ecological chapters have been hurriedly added to revised editions. No biological work is now permitted to conclude without some reference to human ecology and environmental crises.

The purpose of this new ecological series is to survey without undue overlap the major fields of our present environmental confrontation at an introductory college level. The basic text for the series presents an overview of the ecological fundamentals which are relevant to each issue. In association with the works listed in its bibliographical references, it can stand alone as a required text for an introductory college course. For such use each chapter has been provided with a set of review questions. For more extensive courses, the base text leads into each series volume, and the particular area of environmental problems which this explores.

This series treats, subject by subject, the main points of impact in this current ecological confrontation between man and his environment. It presents in breadth and in depth the problems of pollution, pesticides, waste materials, population control, and the resource exploitation which imminently threaten to overwhelm us. Each volume in the series is a definitive study prepared by a specialist in the field, writing from an intimate personal experience of his area, relating but not overlapping his subject with other volumes in the series. Uniquely assembled in each volume will be information which presently is not available without extensive bibliographical research, at the same time arranged and interpreted in a more readily assimilable form. Extensive illustrative material, much of it original, still further facilitates a ready comprehension of the matter presented.

This is an exciting series. The urgency and ferment which have been experienced by all those associated with it cannot fail to be transmitted to the reader. The series confounds the prophets of doom, for it illustrates that given a proper understanding of ourselves and our ecological world, there is yet time for action. This time may be short, but sufficient if we exercise now the characteristics of courage and resolution in which, at times of great crisis, our species has never previously been found wanting.

Arthur S. Boughey

Preface

In recent years, biological control has received considerable attention from persons concerned with the environment. It is not a new subject; indeed it has been the focus of considerable research by pest control experts for some seventy to eighty years. Since, however, there is no brief treatment of the subject currently available that is not highly technical, this book has been prepared as a general introduction for the nonspecialist.

We define biological control as that method of pest control that relies on natural enemies—parasites, predators, and pathogens—to reduce pest populations to tolerable levels. We further define biological control as a natural phenomenon, that component in the control of numbers of any organism, pest or otherwise, which is produced by its natural enemies. This book describes the ecological basis for the phenomenon of biological control, its historical development by man, the nature of entomophagy with particular reference to insects that feed on other insects, procedures for carrying out programs of biological control, and evaluation of any results attained. It goes on to describe factors that limit biological control, to analyze selected case histories where this method of pest control has been attempted, and to show how biological control contributes to integrated control. Although we have limited our coverage mainly to "classical biological control," as defined above, we mention briefly other biological methods of pest control, including host plant resistance, cultural control, the sterile-insect method, and genetic control of pests. The book then ends with a statement about the future of biological control.

Biological control is practiced in a programmatic way, as described in this book, in only a few institutions in America, and indeed in the world. The authors are fortunate in having been associated professionally with this specialty at one of these institutions, the University of California, for more than twenty years. In that period we have participated in most phases of the work, including foreign exploration, mass culture, colonizations, and evaluations, as well as in more or less continuous scientific studies of various phenomena concerned with or underlying biological control. It is from this direct personal experience that we have relied in the writing of this book.

We are pleased to record our debt to the late Professor Emeritus Harry S. Smith, our first "boss" in biological control at the University of California. Harry Smith is recognized as the principal initiator and promoter in America of the biological method of insect pest control in this century. He was the organizer of this specialty in California, first in the state government and later within the California Agricultural Experiment Station. His worldwide influence, enthusiastic personality, and vigorous efforts to develop and teach biological control as a scientific activity are major reasons for the successes it has attained.

We are also indebted to a number of specialists in biological control for advice on one or more aspects covered in the book. We particularly wish to acknowledge the helpful advice of C. B. Huffaker, K. S. Hagen, L. E. Caltagirone, all of the Division of Biological Control, University of California, Berkeley, and L. A. Andres, Biological Control of Weeds Investigations, A.R.S., U.S.D.A., Albany, California. We are also indebted to F. J. Simmonds, Commonwealth Institute for Biological Control for advice concerning certain aspects of our treatment of the history and development of biological control.

1

The Nature and Scope of Biological Control

Biological control is a natural phenomenon—the regulation of plant and animal numbers by natural enemies (biotic mortality agents). It is a major element of that force, natural control, which keeps all living creatures (excepting possibly man) in a state of balance.

This volume is essentially concerned with the biological control of insects and weedy plants by one group of animals, the insects. In a sense, then, the title is inexact, since the full scope of biological control is not covered. But this should not be a major concern because the basic principles of the phenomenon are the same for all groups, and in the applied area of biological control, the overwhelming emphasis has been on pest insects and weeds, with insects the principal biological control agents involved. (See Figure 1.)

This book deals with those biotic agents that prevent the normal tendency of populations of organisms to grow in exponential fashion and with the mechanisms by which such growth is prevented. To gain some insight into the awesome significance of uninhibited multiplication of an organism we have only to look to *Homo sapiens* to see the effects of our population explosion on the environment in the way of water and air pollution, decimation of plant and animal life, destruction of soil fertility, and so on. Since insects comprise an estimated 80 percent (perhaps 1 to 1½ million species) of all terrestrial animals, even the partial inhibition of naturally occurring biological control would engender unimaginable consequences. Man might not survive the intensive competition for food and fiber he would face or the reduced health he would suffer from the unleashed hordes of insects.

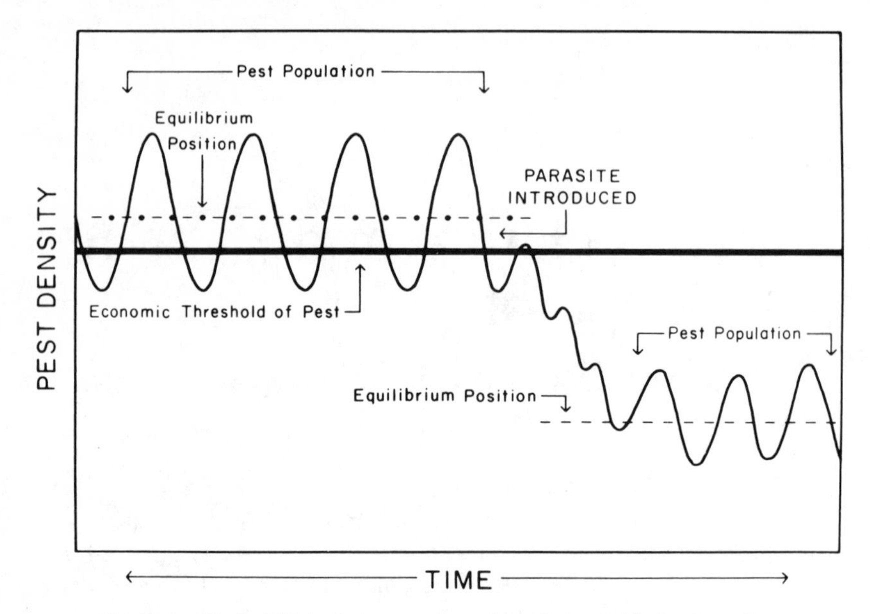

Figure 1. Classical biological control resulting in total elimination of an insect pest as an economic problem. Note that it is not the economic threshhold (an artifact of man) that is affected by the introduced parasite, but rather the pests' equilibrium position (long-term mean density). From Smith and van den Bosch 1967, with permission of the publisher.

Biological control, then, is of great importance to us and most probably critical to our survival.

DEFINITION OF BIOLOGICAL CONTROL

DeBach (1964) discussed the semantics of the term *biological control.* He concluded that the term can refer to a natural phenomenon, a field of study, or an applied pest control technique involving manipulation of natural enemies. In this light, the following definition seems most appropriate since it is simple and yet embraces DeBach's three semantic variations. Biological control is thus defined as *"the action of parasites, predators and pathogens in maintaining another organism's density at a lower average than would occur in their absence"* (DeBach 1964). Close analysis of these few words reveals that they describe a natural phenomenon, denote a field of study, and accommodate the possibility of deliberate natural enemy manipulation. We would find it very difficult to concoct a better definition of biological control.

Some people hold a wider view of biological control that embraces such factors as host resistance, autosterilization, and genetic manipulation of species. But we prefer the narrower concept, first because it is the traditional one, and second because it is neatly delimited by the phenomena of predation, parasitism, and pathogenesis.[1] However, to provide insight into their variety and nature, Chapter 9 treats the other biotic methods of pest control.

A COMPARISON OF THE BIOLOGICAL CONTROL OF PEST INSECTS AND WEEDS

In principle there is little difference between the biological control of insects and weeds. Both involve natural enemies which act to suppress or maintain pest or potential pest species below economically injurious levels. And with each, where natural enemy importation has been employed, the successes have been overwhelmingly against alien pest species. But there are some differences in the biological control of the two pest groups. For one thing, with the plant-feeding insect, a high degree of host specificity, preferably monophagy, is an absolute necessity, for there cannot be the remotest chance that the species will develop an affinity for any plant of economic value. Therefore, insects under consideration for importation against weedy plants are subjected to intensive feeding and host preference tests before being cleared for release. There is simply no margin for error in this process, for once a weed-feeding insect is released into the new environment it cannot be called back. On the other hand, with entomophagous insects, oligophagy or even polyphagy may sometimes be advantageous, and certainly there is no hard and fast stipulation that an imported parasite or predator be narrowly specific. With entomophagous insects the basic concerns are merely that no beneficial species (e.g., the honeybee, lady beetles) be endangered or that hyperparasites (parasites of parasites) be imported.

---

1 Entomologists customarily use the term *parasite* for insects which are parasitic in or on other insects, and the term *pathogen* for microorganisms which cause disease in insects. On the other hand, parasitologists concerned with the medical and veterinary sciences commonly use the term *parasite* to refer to any organism which lives on or in a host, including both microbes and multicellular organisms. To avoid confusion, entomologists often distinguish parasitic insects as *parasitoids*. The distinction is elaborated further in the text. Where we use the terms *parasite, parasitism,* and *parasitic,* these are to be understood as referring to insects living in or on host insects. Parasitic microbes which attack and cause disease in insects are called *pathogens*.

The need for specificity in weed-feeding insects places a heavy burden of responsibility on everyone involved in the importation process, particularly those who do the actual testing. In practice, the candidate insect species is first intensively tested in the overseas collecting area for biological characteristics, host plant affinities, and oviposition habits. The plants used in the feeding tests range from wild species related to the host weed through a variety of plants of economic value.

The overseas testing establishes whether an insect species will be passed on to the domestic quarantine facility for additional intensive study and testing. Then, after this second screening is completed, the data are reviewed by a special committee of experts who make the final decision as to whether field releases shall be made.

The suppression of weedy plants by imported natural enemies differs somewhat from the suppression of insect pests by natural enemies. With insect pests, suppression usually results directly from premature mortality produced by the natural enemy. But with weeds, the role of the natural enemy is more complex; thus (1) it may directly "kill out" its host, (2) it may so weaken or stress the weed that aggressive competitors displace it or make it susceptible to other, preexisting mortality agents, (3) it may impair the reproductive capacity of the weed by destroying its seeds or flowering parts, or (4) its feeding lesions may create an avenue for fatal infection by pathogens. For an example, see Figure 2.

But despite the technical and subtle mechanical differences just described, biological control of insects and weeds operates under the same broad principles, and the two will be treated together in the following pages.

INSECT PATHOGENS

Instead of dealing with insect pathology in detail, this book places its major emphasis on entomological predators and parasites of insects and weeds. However, pathogens are biological control agents and it would be a serious oversight not to attempt to place them in proper perspective as regards biological control.

Pathogenic viruses, bacteria, fungi, protozoa, and nematodes play important roles in the regulation of plant and insect numbers (Steinhaus 1949, 1963). A wide range of pest and potential pest species are afflicted by diseases, which either prevent them from attaining damaging levels or greatly reduce their potential to cause injury. Among the major pest insect groups, species of Orthoptera, Homoptera, Hemiptera, Lepidoptera, Coleoptera, Diptera, Hymenoptera, and even species in the Acarina (e.g., Tetranychidae)

Figure 2. Biological control of Alligatorweed; A, The leaf beetle *Agasicles hygrophila* Selman and Vogt; B, Weed infestation in the Ortega River, Jacksonville, Florida, November 1965 before release of *Agasicles;* C, Same site 15 months after beetle release. Photograph A by R. D. Hennessey. Photographs B and C by C. F. Zeiger

suffer significantly from disease. In other words, microbial parasites are frequently major contributors to naturally occurring biological control.

Man has long recognized the important role of pathogens as controlling agents in nature, and his thoughts long ago turned to ways to deliberately manipulate them. For example, d'Herelle, early in this century, deliberately distributed the bacterium *Coccobacillus acridiorum* d'Herelle, in an attempt to initiate epizootics in grasshopper populations. At about the same time there was also an attempt to induce epizootics of the fungus *Beauveria bassiana* Glover in populations of the chinch bug, *Blissus leucopterus* (Say), in the midwestern United States. Neither attempt was considered successful, but they were precursors of later successful programs.

Deliberate introductions of exotic pathogens have not been so widely attempted as have introductions of parasitic and predaceous insects. But there have been some successes in pathogen introductions such as that of the polyhedrosis virus of European sawfly, *Diprion hercyniae* (Hartig), from mainland Canada into Newfoundland, and that of the granulosis virus of codling moth, *Laspeyresia pomonella* (Linnaeus), from Mexico into California.

But the greatest strides in the manipulation of pathogens have been in their development as microbial insecticides. The first pathogen to be developed, marketed, and utilized in this way was *Bacillus popilliae* Dutky, the famed milky disease of the Japanese beetle, *Popillia japonica* Newman. An even more outstanding microbial insecticide has been *Bacillus thuringiensis* Berliner, which in its various strains is an effective killer of a broad spectrum of lepidopterous pests. *B. thuringiensis* is under commercial production both in the United States and abroad, and must be considered a major modern insecticide. It is a particularly desirable material because of its high degree of selectivity among arthropods and its complete safety to the environment and warm-blooded animals.

Several viruses have also been effectively used as microbial insecticides. The viruses are even more selective than *B. thuringiensis* and thus would seem to be of high promise as safe insecticides. However, since they are viruses, they are being exhaustively tested for possible pathogenicity to other groups of animals. This has slowed their federal registration for insecticidal use, increased their developmental costs, and delayed their commercial exploitation. However, it seems only a matter of time before the registration protocol is developed for viruses permitting such promising ones as the polyhedrosis viruses of bollworm (*Heliothis zea* [Boddie]), cabbage looper (*Trichoplusia ni* [Hübner]), and beet armyworm (*Spodoptera exigua* [Hübner]) and the granulosis virus of codling moth (*L. pomonella*) to be brought into widespread use.

REVIEW AND RESEARCH QUESTIONS

1. What is the definition of biological control?

2. List the three different classes of biological control agents used to control insect pests.

3. Describe the special features which are one must be concerned with in the biological control of weeds.

4. Discuss the history and importance of insect pathogens in relation to biological control and give two examples of recent successes.

5. What is a microbial insecticide?

BIBLIOGRAPHY

Literature cited

DeBach, P., ed. 1964. *Biological control of insect pests and weeds.* London: Chapman & Hall, 844 pp.

Steinhaus, E. A. 1949. *Principles of insect pathology.* New York: McGraw-Hill, 757 pp.

Steinhaus, E. A., ed. 1963. *Insect pathology—an advanced treatise.* New York and London: Academic Press, Vol. I, 661 pp., Vol. II, 689 pp.

Stern, V. M., R. F. Smith, R. van den Bosch, and K. S. Hagen. 1959. The integration of chemical and biological control of the spotted alfalfa aphid. Part 1. The integrated control concept. *Hilgardia* 29:81-101.

Additional references

Clausen, C. P. 1940. *Entomophagous insects.* New York: McGraw-Hill, 688 pp.

Huffaker, C. B., ed. 1971. *Biological control.* New York: Plenum, 511 pp.

National Academy of Sciences. 1968. The biological control of weeds in *Principles of plant and animal pest control,* Vol. II, Chap. 6, pp. 86-119. Nat. Acad. of Sci. Publ. 1597, 471 pp.

Sweetman, H. L. 1958. *The principles of biological control.* Dubuque, Iowa: Wm. C. Brown, 560 pp.

van den Bosch, R. 1971. Biological control of insects. *Ann. Rev. Ecol. and Systematics* II: 45-66.

2 Ecological Basis for Biological Control

Biological control is a natural ecological phenomenon which, when applied successfully to a pest control problem, can provide a relatively permanent, harmonious, and economical solution. But, because biological control is a manifestation of the natural association of different kinds of living organisms, i.e., parasites and pathogens with their hosts, and predators with their prey, the phenomenon is a dynamic one, subject to disturbances by other factors, to changes in the environment, and to the adaptations, properties, and limitations of the organisms involved in each case (Huffaker and Messenger 1964).

In order to understand the potential, and the limitations, of biological control, and especially in order to carry out competently a program of biological control, an awareness of the ecological basis of the phenomenon is essential. Three interrelated concepts which must be understood are (1) the idea of discrete populations and communities, (2) the balance of nature, and (3) the natural control of numbers.

Biological control is a manifestation of the association of different, interdependent species in nature. But species exist as groups of like individuals. These interbreed, reproduce, and die. By reproducing they maintain themselves as a group, which at a local level is called a *population* (Boughey 1971).

A population changes in size, that is, in the number of individuals it contains, according to whether environmental circumstances favor the production of more, or less, than the number of individuals dying in a given interval of time. Populations are thus dynamic with regard to size.

Another important feature of a population is its age structure. *Population age structure,* a term often simplified to population structure, means the distribution age pattern of the individuals in the population. The population structure may appear in one of two extremes, or in patterns intermediate between these extremes. In one extreme all members of the population at any one time are approximately the same age or are in the same stage of development. In the other extreme individuals of all ages occur together. In the first case the life cycles of all population members are synchronized with each other, a situation often brought about by annual climatic cycles or by producing only one generation a year. In the second case generations are not synchronized, but rather they overlap, a pattern commonly found in populations of short-lived insects with many generations per year or in populations of insects which display continuous reproductive activity uninterrupted by seasonal climatic cycles.

Population structure is important in respect to host populations in which only one or two stages of development are utilizable by a particular parasite species. Close synchronization between parasite and host life cycles must occur if the parasite is to be successful in controlling the host. Such is the reason why the aphelinid parasite *Metaphycus helvolus* (Compere) is an effective control agent of citrus black scale, *Saissetia oleae,* (Bernardi), in coastal southern California, while it is much less effective against the same pest in interior southern or central California. In the former area the scale population structure includes all ages at once because of the lack of synchrony in the life cycle; in the latter areas the scale population structure is restricted to only one or two age classes at any given time, so that for certain periods of the year no suitable host stages are available for the parasite to attack. The parasite population is then unable to reproduce and diminishes in numbers as a result.

Population age structure is also important in the ecology of an insect since it often reflects the growth phase of the population. When a population is young and is beginning to increase in numbers, its age structure includes relatively many of the younger ages, less of the intermediate ages, and few adults. When a population is mature and no longer increasing in numbers, perhaps because of intraspecific crowding, it is then composed of relatively less young and more adults. A population which is declining because of overcrowding contains relatively few young and many adults.

Populations are also dynamic with regard to geographic distribution. They tend to spread in space until some limiting environmental condition is encountered, such as a geographic barrier like a coast, or mountain range, or desert boundary, or an environmental limitation such as the absence of a required resource, like a necessary food organism, or a soil habitat.

Populations do not exist in isolation; they occur in habitats in association with other species. Such assemblages of species populations constitute *communities.* To the degree that certain species are consistently associated together and can be recognized, particularly certain characteristic plants, we can distinguish discrete communities (Odum 1971). For example, we can describe a pine forest community, and can usually find certain species of insects and other animals associated with the dominant pine species.

In communities we can distinguish trophic or nutritional associations between interacting species. Thus we recognize primary producers, or plants, primary consumers, or herbivores, secondary consumers, or carnivores, decomposers, and scavengers, and so on (Odum 1971). *Food chains* can commonly be discerned, wherein a given plant species is consistently fed on by a defoliating insect, for example, which in turn is fed upon by certain bird species. This is a three-step chain; where the insectivorous bird population is preyed upon by a hawk we recognize a fourth trophic step. Such food chains, because of simple material and energy losses along the way, are not endless, but typically occur in links of four to six.

Where food chains branch or join together, as generally happens in complex communities, the complex of trophic paths is referred to as a *food web.* Such webs can be discerned where one herbivore feeds on more than one plant species, or where several bird species include one defoliating insect in their diet.

Thus we encounter on a universal scale such trophic interactions as *phytophagy,* or the consumption of plants or plant parts by herbivorous animals, and *carnivory,* or the consumption by animals of other animals. In biological control of pest insects we are particularly interested in *entomophagy,* the consumption or nutritional dependency of certain species of animals on insects, particularly phytophagous ones. Thus we work with such entomophagous animals as insect-eating mammals (shrews, mice), insectivorous birds and amphibians, and, particularly, entomophagous insects.

It is customary in biological control work to describe the species of animals and plants which live at the expense of other plants and animals as natural enemies of the latter. Any given species in a community, with few exceptions, is attacked and fed on by one or more such natural enemies, an indication of the tremendous potential for biological control.

All organisms are capable of increasing in numbers through processes of reproduction. Most insects are particularly notable for their high potential rates of numerical increase because of their relatively very high fecundities and short life cycles. But the fact is that most organisms, including the rapidly breeding insects, do not increase over successive generations or for prolonged periods. On the contrary, they increase only periodically and to limited extent, as a consequence of natural controls present in their environments (Solomon 1949).

These natural controls generally limit the numbers of insects. Such checks to numerical growth include limited resources (food, space, shelter), periodically occurring inclement weather or other hazards (heat, cold, wind, drought, rain), competition from themselves or from other kinds of animals, and natural enemies (predators, parasites, pathogens). (See Figure 3.) This last category is particularly important for many insect species, for while resources may rarely appear to be in short supply, weather can be constantly favorable, and competitors scarce or absent, natural enemies are almost universally present, often significantly so.

Probably every insect population in nature is attacked to some degree by one or more natural enemies. Indeed, referring just to entomophagous insects alone, Clausen (1940) has stated that probably every phytophagous insect species is attacked by one or more parasitic or predatory insects. Other predatory animals can also act as natural control agents of insect populations, for example some birds, certain mammals, toads, frogs, and lizards have this role.

The result of natural control is the regulation of numbers, preventing the population from becoming too high or relaxing certain suppressive influences when the population becomes low. The occurrence of this situation, the long-term maintenance of the population at a characteristic level of abundance relative to other organisms in the community, is the demonstration of the *balance of nature.* The mechanisms and interactions between the population and its environment which bring about the relative balance of numbers constitute the natural control of populations. Natural control includes the collective forces of the environment which serve to hold a given population in check against its own capability for numerical growth. As such, natural control, especially through mortality factors or influences, but to a lesser degree also through factors acting on natality or reproductive capability, includes climatic factors, such as excessive heat or cold, or aridity, or disappearance or deterioration of food resources, the action of competing species, and natural enemies.

It is useful to distinguish between environmental factors which act as mortality agents at intensities unaffected by the size of the population, for example, weather, and those whose intensities of action vary with the abundance of the species in question, such as predators which consume proportionately more prey when the latter are abundant than when they are scarce. (See Figure 3.) The former type of limiting factor is called a *density-independent* mortality factor, and the latter type is called a *density-dependent* mortality factor (Smith 1935). Natural enemies have the capability of acting as density-dependent mortality factors, though they may not act this way, depending on their own environmental limitations. All

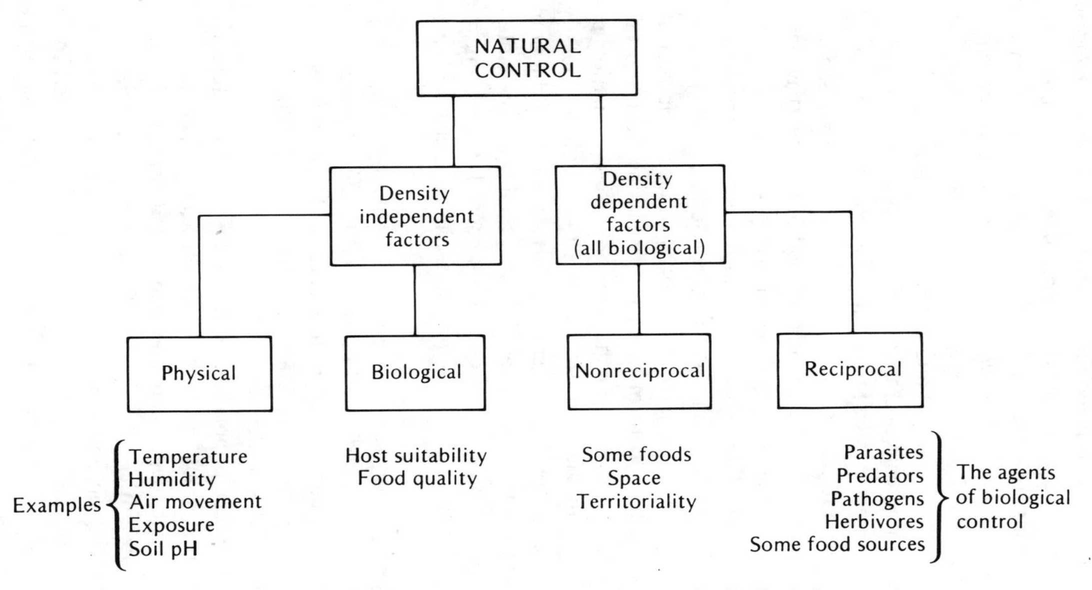

Figure 3. The major components of the natural control of population numbers.

natural enemies used in or contributing to successful biological control programs act in this way.

Density-dependent factors may also be classified according to whether they vary in numbers (or magnitude) as the host numbers change (*reciprocal action*) or whether their numbers (or magnitude) remain fixed (*nonreciprocal action*). Parasitoids and predators are examples of the former, since they commonly increase in numbers when their hosts or prey become numerous, and decrease as the hosts or prey become suppressed. That is to say, the enemies "control" their hosts, and the hosts "control" the enemies. Space is an example of a nonreciprocal factor since it does not wax and wane as the user population rises or falls. Space can "control" the numbers of the users, but the users do not alter the amount of space present. Some herbivorous insects can be limited by the amount of food (host plants) available to them, but often do not influence the number of plants present.

For insect populations to remain in relative numerical balance within their normal communities for substantial periods of time, it appears to us necessary that there be one or more density-dependent mortality agents involved in the natural control of such populations. Such mortality agents, being responsive to increases in the density of the population in question, serve as regulators to check this increase; as the population density declines, the regulative action of these agents moderates, allowing the population to rise again. Using control-system terminology, the insect population density is regulated by means of a negative-feedback control mechanism, much like a governor on a steam engine or a thermostat in a heating system (Nicholson 1954).

The evidence for density-dependent regulation of insect population densities comes from a variety of sources, but for our purposes it is sufficient to emphasize the occurrence of biological control both in undisturbed nature and in highly managed agricultural environments (agroecosystems), as demonstrations of this contention.

When an insect population is maintained at a characteristic level of abundance by the effects of natural control agents (including all mortality agents, both density-dependent and density-independent in action), there usually is a substantial contribution from natural enemies to the total mortality occurring in any generation. Furthermore, the great majority of insects are rare—of the 1 million or more insect species, perhaps only from 10,000 to 30,000 of those which feed on crops or livestock or menace our health or possessions are abundant enough to be recognized as economically important. Therefore, when insect species invade new geographic regions, as an accidental result of man's commercial activities, for example, they many times increase to extraordinarily high numbers mainly because they have escaped the controlling influences of their customary natural enemies. Overall generation mortality is greatly diminished, while the reproductive capacity of the species remains high. Populations of such invaders therefore begin to increase in number at an exponential rate. We soon have in such cases a population outbreak.

When such an invading insect is injurious, pest outbreaks occur, and control efforts must be brought to bear. It is very logical, then, that with such pests efforts be directed to the search for and colonization of any adapted natural enemies that remain behind in the native home of the invading species. Virtually all successful classical biological control programs to date have resulted from the reassociation of invading pests of foreign origin with their adapted natural enemies (DeBach 1964).

However, not all insect pests are of foreign origin; some, including a number of the most serious, are native species. The very important cotton bollworm, *Heliothis zea,* and the devastating lygus bug, *Lygus hesperus* Knight, are examples of pests native to the western United States, as are the apple maggot, *Rhagoletis pomonella* (Walsh), and the spruce budworm, *Choristoneura fumiferana* (Clemens), to the northeastern and north central parts of the United States and Canada. The question is, can biological control provide any help against such pests as these?

The case for biological control of native pest species is technically more complicated than it is for the invading, foreign pest. A native species must be presumed to have already associated with it an assortment of adapted natural enemies, and after investigation this is usually found to be the situation. But the agricultural practices of the growers of the crop under attack by the native pest tend in many instances to favor unduly the increase in numbers of the pest. Such practices also often interfere with the efficacy with which native natural enemies exert their share of the natural control of such pests. Biological control of such native pests can then assume several routes: (1) the introduction of natural enemies of foreign origin which are associated with related pest species, (2) the modification of agricultural and other practices with the intention of enhancing native natural enemy action, or (3) the employment of other pest control techniques, chemical control in particular, to bring about the integrated control of such pests.

Some invading pests have been controlled by natural enemies derived from related host species. For example, in Hawaii the importation of the Queensland fruit fly parasite, *Opius tryoni* Cameron, provided an important biological control agent against the Mediterranean fruit fly, *Ceratitis capitata* (Wiedemann). This has led to hope that native pests might be similarly controlled. However, parasites, which experience shows are most often the best biological agents, are in the great majority of cases highly host specific.

That is, their natural adaptations to the parasitic mode of life are narrowly limited to the particular host species with which they have evolved. Past efforts to use imported parasites from related hosts on native host species have only rarely succeeded. This is not true for predators. However, the knowledge that some parasites can attack more than one host species indicates that this possibility should not be neglected in development of future biological control programs.

Alteration of normal agricultural practices in some cases has favored the increase of native natural enemies. The strip-cropping of alfalfa protects the numerous beneficial insect species associated with this plant, resulting in better control of the phytophagous pest species. The addition of shelter boxes to tobacco fields increases the numbers of predatory wasps which help control tobacco worms. The planting of wild blackberry plants near grape vineyards provides an alternate host for an egg parasite of the grape leafhopper.

The inclusion of biological control agents in the integrated control programs against some of our most serious native pests has proven to be of such value, but also of such complexity, that we present a fuller treatment of integrated control later in this book.

REVIEW AND RESEARCH QUESTIONS

1. What are the three ecological bases for biological control?

2. Discuss the position of biological control agents in the trophic structure of communities.

3. How are density-dependent mortality factors related to population regulation?

4. How does the biological control of native pests differ from that of invader pests?

5. Describe some of the ways in which natural enemies may be augmented to bring about biological control.

BIBLIOGRAPHY

Literature cited

Boughey, A. S. 1971. *Fundamental ecology.* New York: Intext Educational Publishers, 222 pp.

Clausen, C. P. 1940. *Entomophagous insects.* New York: McGraw-Hill, 688 pp.

DeBach P. ed. *Biological control of insect pests and weeds.* London: Chapman & Hall, 844 pp.

Huffaker, C. B., and P. S. Messenger. 1964. The concept and significance of natural control. In *Biological control of insect pests and weeds* ed. P. DeBach, Chap. 4, pp. 74-117.

Nicholson, A. J. 1954. An outline of the dynamics of animal populations. *Austral. J. Zool.* 2: 9-65.

Odum, E. P. 1971. *Fundamentals of ecology.* 3rd ed. Philadelphia: Saunders, 574 pp.

Smith, H. S. 1935. The role of biotic factors in the determination of population densities. *J. Econ. Ent.* 28: 873-898.

Solomon, M. E. 1949. The natural control of animal populations. *J. Anim. Ecol.* 18: 1-35.

Additional references

Allee, W. C., A. E. Emerson, O. Park, T. Park, and K. P. Schmidt, 1949. *Principles of animal ecology.* Philadelphia: Saunders, 837 pp.

Solomon, M. E. 1969. *Population dynamics.* New York: St. Martin's, 60 pp.

Whittaker, R. H. 1970. *Communities and ecosystems.* New York: Macmillan, 158 pp.

3 The History and Development of Biological Control

The purposeful control of insect and weed pests by biotic agents is a comparatively modern development, having become an effective technique in pest control only since about 1890. However, there are antecedent, historical events which trace the evolution of some of the fundamental concepts in the development of biological control, and several of these events show the remarkable and perceptive insight of man into the workings of nature. Without these prenineteenth-century discoveries and conceptualizations, modern environmental science, to which biological control has made substantial contributions, would very likely have been much delayed. These discoveries and concepts include, among others, that of the balance of nature, population growth and limitation, natural control of numbers, the symbioses among different species, particularly those of plants, animals and their natural enemies, and the roles such natural enemies play in the determination of abundance.

The history of applied biological control, therefore, in no little degree reflects our increasing knowledge of ecology. Indeed, particularly during the late nineteenth century, the ideas and concepts underlying biological control contributed in important ways to the developing theories and principles of ecology. Reciprocally, the emergent general theories and principles of ecology reinforced the practical formulation of biological control of pests. This is not surprising, since biological control is in its essence an ecological phenomenon, and in its practice is an example of "applied ecology."

A number of books on entomology or biological control include useful sections on the historical features of this subject. These include, in chronological order, *A History of Entomology* by E. O. Essig (1931), *The Biological Control of Insects* by H. L. Sweetman (1936), *The Principles of Biological*

Control by the same author (1958), *Beneficial Insects* by L. A. Swan (1964), and the very comprehensive *Biological Control of Insect Pests and Weeds,* edited by P. DeBach (1964). Besides consulting much of the original literature, we have gleaned some of the following material from one or more of these sources, leaning heavily on the detailed chapter called "The Historical Development of Biological Control" prepared by our colleague R. L. Doutt (1964), contained in the collection mentioned above edited by DeBach. We must acknowledge two further references, particularly for facts concerning details of the natural enemies used or of the projects carried out in the United States since the beginning of the twentieth century. These are *Entomophagous Insects* (1940) and *Biological Control of Insect Pests in the Continental United States* (1956), both by C. P. Clausen.

ANCIENT ORIGINS

The idea that insects could be used intentionally to suppress populations of other insects is an ancient one. Our earliest knowledge of this practice seems to have originated with the Chinese, and, not surprisingly, involved the use of predators, in this case predatory ants, to control certain insect pests of citrus. Indeed, this agricultural practice has descended through the ages, continuing even into modern times in the Orient where citrus growers maintain, and sometimes even purchase, colonies of the predatory ant *Oecophylla smaragdina* Fabricius, to place in orange trees to reduce the numbers of leaf-feeding insects (McCook 1882, Clausen 1956).

This use of predatory ants no doubt derived from the obviousness of the carnivorous habit in this group of insects. Our ancestral farmers must have been aware of the foraging behavior of such forms as the voracious army ant and the carrying-off of soft-bodied grubs and caterpillars by the ubiquitous trail-making species of *Lasius* and *Formica.* But the practical use of ants by the ancient agriculturalists to control pests was not accomplished without some considerable ingenuity. Not all ants are predatory. Indeed, medieval date growers in Arabia seasonally transported cultures of predatory ants from nearby mountains, where presumably they occurred naturally, to the oases to control *phytophagous ants* which attacked date palm. This practice constitutes the first known example of the movement, by man, of natural enemies for purposes of biological control. It also bears testimony to the ability of the medieval Arabian date growers to distinguish among the species of the ant family on the basis of their food habits. And, from what we know today about the habits of certain predatory ants, for instance the Argentine ant, *Iridomyrmex humilis* Mayr, which actually protect certain kinds of phytophagous insects, such as aphids, soft scales, and mealybugs, from their natural enemies, we must admire still further the insight displayed by these early practitioners of biological pest control.

Awareness of the possibilities for use of parasitic insects in combating pests was slower in development, most likely because of the much more subtle and cryptic nature of the parasitic habit in insects. Insect parasitism was first recognized by Vallisnieri (1661-1730), who, in Italy, noted the unique association between the parasitic wasp *Apanteles glomeratus* (Linnaeus) and the cabbage butterfly *Pieris rapae* (Linnaeus) (Doutt 1964). In the early decades of the eighteenth century there appeared an increasing number of reports referring to the parasitic habit among insects, but the idea that such natural enemies could be used in a practical sense to control pests was delayed until the following century.

The first suggestions that biological control through parasites might be a practical solution to insect pest problems were European in origin. Erasmus Darwin (1800) noted the destruction of cabbage butterfly caterpillar infestations by the "small ichneumon-fly which deposits its own eggs in their backs" (Doutt 1964). The gathering and storing of parasitized caterpillars in order to harvest parasite adults for later release was proposed by Hartig in 1827 in Germany (Sweetman 1936). In France, Boisgiraud in 1840 collected and liberated large numbers of predatory carabid beetles, *Calasoma sycophanta* (Linnaeus), to destroy leaf-feeding larvae of the now famous gypsy moth. In Italy, Villa in 1844 proposed, and later demonstrated, the use of predatory insects, in his case carabid and staphylinid beetles, to destroy garden pest insects.

In Europe, the initial applications of biological control mainly concerned the use of locally derived, normally present parasites and predators to control local infestations of pests. There was no suggestion or effort to bring such natural enemies from distant places to control local pests. This was very likely because most European agricultural or garden pests were considered native to the localities infested, crops were rarely ravaged by mass outbreaks of pests of foreign origin, and the idea that insect parasites and predators from foreign lands might prove useful in combating pest infestations probably never came to mind.

NORTH AMERICAN BEGINNINGS

In America, with the development and rapid spread of agriculture from the eastern seaboard westward with the expanding frontier in the early nineteenth century came the increasing occurrence of insect pest outbreaks resulting mainly from insect species of foreign origin. For example, the highly damaging wheat midge *Sitodiplosis mosellana* (Gehin) was known to have come to America from Europe, and Asa Fitch (1809-1879), State Entomologist for New York, conjectured that its persistence at injurious levels was due to the absence of its normally restrictive natural enemies which in Europe serve to keep it at lower, less harmful numbers. Fitch proposed in 1855 the importation of parasites from England to bring about a reduction in abundance of the midge. While this pioneering proposal came to no immediate practical end, the idea soon received support from other foresighted entomologists, including C. J. S. Bethune in Canada and Benjamin Walsh (1808-1870) in Illinois.

Some years later, in 1870, parasites of the plum curculio *Conotrachelus nenuphar* (Herbst) a native pest species, were distributed from one part of Missouri to another for control purposes by State Entomologist C. V. Riley (1843-1898). In 1873 Riley arranged the first international shipment of a natural enemy in the transfer of the predatory mite *Tyroglyphus phylloxerae* Riley to France from North America for possible control of the grape phylloxera, a native North American pest accidentally introduced into Europe in the early nineteenth century.

In 1879 Riley was appointed Chief Entomologist for the U.S. Department of Agriculture, Washington, D.C. Soon thereafter (1883), he directed the importation of the internal parasite of the cabbage butterfly, from England to America. This introduction was successful, and *A. glomeratus* eventually became well distributed throughout the eastern and middle western states. However, it did not become a very effective biological control agent, and the popularity of the method remained at low ebb.

California origins

So far, the scene for the development of biological control in America remained eastern and midwestern. But, in the early 1880s, the "movement" found support in the burgeoning agricultural enterprises of California. During the period between 1840 and 1870, California agriculture, under the stimulus of a mild climate and exceptionally fertile soils, expanded at an extraordinary rate. A multiplicity of new crops, resulting from importation of seeds, seedlings, and cuttings of the best varieties of fruit and nut trees, vines, field crops, and ornamental plants from many countries, enabled California to become one of the world's major crop producing regions. But unfortunately, along with the importation of all these plants came an increase in insect and disease problems. It is therefore not surprising that as early as 1881 proposals for introducing natural enemies for pest control purposes were submitted by several California horticulturalists. But it was not until the late 1880s that the first planned, successful project in biological control took place, involving the serious citrus pest known as the cottony-cushion scale, *Icerya purchasi* (Maskell).

The cottony-cushion scale control campaign

The biological control project against the infamous cottony-cushion scale in California constitutes not only the first truly successful example of the use of this pest control technique in the world, but also it serves as a classic example, exhibiting all of the basic features characteristic of this method of pest control. The project is worth exploring in some detail.

The cottony-cushion scale, a destructive pest of citrus, pear, acacia, and other plants, was first recorded from California in the town of Menlo Park in 1868, where it was noted infesting acacia plantings in a horticultural nursery. From the original nursery infestation it soon spread to nearby ornamentals, including citrus. At this time in California the young, burgeoning citrus industry was concentrated mainly in the Los Angeles area some 400 miles south of Menlo Park. But within three or four years of its original invasion in the Bay Area the pest was unwittingly carried to southern California on infested lemon stock. It became established in the Los Angeles area prior to 1876, for by this time it was discovered well spread through the various citrus groves of what is now the downtown area of that city. Soon thereafter it was found in San Gabriel Valley citrus groves, some 10 miles to the east, and in the newly developing groves of Santa Barbara, 100 miles to the northwest. By 1880 the pest had become distributed throughout California, damaging citrus trees wherever they were grown.

Specimens of the cottony-cushion scale were sent to C. V. Riley in 1872, while he was still State Entomologist for Missouri. He suggested that the new pest may have come from Australia, since he knew that much citrus nursery stock was being imported from the Orient and the South Pacific at that time, and because the pest resembled closely a scale pest of that country known as the dorthezia. Then, shortly after becoming Chief Entomologist for the federal government, Riley traveled to California to observe major pest problems there, and took particular note again of the increasingly damaging scale pest. Because the cottony-cushion scale had first been described by a New Zealand entomologist, W. M. Maskell, Riley reiterated that the source of the invading pest must be the Australasian region. This assumption was soon confirmed by Maskell, who, in correspondence with W. G. Klee in San Francisco, stated that *I. purchasi* was indeed native to Australia.

In the middle 1880s, both Riley and Klee wrote about the possibilities of importing beneficial insects to control the cottony-cushion scale. Klee wrote to the Australian entomologist Frazer Crawford to inquire about such natural enemies, and received positive information about several possibilities. After some effort at procuring funds to support a natural enemy search, Riley, with the moral support of the California State Board of Horticulture, was able to assign U.S. Department of Agriculture entomologist Albert Koebele, then stationed in California and familiar with the scale pest, to this overseas venture. Koebele's mission took place in 1888.

Soon after arriving in Australia, Koebele found two enemies attacking the cottony-cushion scale on citrus, one a dipterous parasite, *Cryptochetum iceryae* (Williston). The other was a coccinellid predator (lady beetle), commonly known as the Vedalia, whose scientific name is *Rodolia cardinalis* (Mulsant). Both were sent by ship to San Francisco where they were examined, reared, and released in Los Angeles as adults on scale-infested citrus trees enclosed in canvas tents. The lady beetles immediately began feeding and ovipositing on scale infestations and rapidly increased in abundance. They were thereon allowed to spread into adjacent trees.

The results of these colonizations were dramatic. The lady beetles multiplied and spread rapidly. So did the other natural enemy, the little parasitic fly *Crypotochetum*, which had been colonized along with the Vedalia, though it received little publicity, perhaps because of its less visible habits. Scale infestations were reduced sharply, and within months the Southern California cottony-cushion scale epidemic had been reduced to harmless levels. (See Figure 4.)

The post-vedalia expansion of biological control

The marked success of the cottony-cushion scale project in California and its extension to many other parts of the world, coupled with its permanency, simplicity, and cheapness led to the enthusiastic support of similar ventures toward the solution of other agricultural pest problems. The method was envisioned as the utopian answer to age-old insect plague problems which have afflicted man throughout his history. Albert Koebele, who returned to the United States early in 1889, went back to Australia to look for more natural enemies of insect pests of concern to California. On this second mission (1892), Koebele concentrated mainly on predatory species, and the great majority of the enemies found and shipped back to California on this second trip were coccinellids. This was very likely because of the impressive success of the previous discovery, the Vedalia beetle. Thus he sent to California the lady beetle *Cryptolaemus montrouzieri* (Mulsant), a natural enemy of another group of serious citrus pests, the mealybugs. This predator, soon given the name mealybug destroyer, became well established for a time in Southern California, particularly along the coast where it was seen to be an

Figure 4. A vedalia beetle, *Rodolia cardinalis,* feeding on a cottony-cushion scale.

avid attacker of mealybugs on citrus. But any potential for continuous control was noted to be inhibited by its sensitivity to winter conditions each year. This maladaptedness led to reviving the idea of artificial propagation of the beetle in insectaries (an idea proposed by Felix Gillet (1881), an early California horticulturalist), with the release of large numbers in infested groves each spring and summer. This technique, now named *periodic colonization,* was put into effect in 1919 with considerable success.

One of the other more or less effective predators, *Rhizobius ventralis* (Erichson), introduced into California in 1892, holds historical interest in another way relative to biological control. A predator of black scale, it seems to have been involved in what may be the first reported observation of the interference of natural enemy action by pesticides. In 1893, in Santa Barbara, where kerosene emulsion was used as a spray for controlling black scale on olive trees, it was noted that where sprays were used, scale infestations remained abundant. In such groves, no *R. ventralis* were to be found. On the other hand, in nearby unsprayed olive trees, the scale was observed to be less abundant, and *R. ventralis* was much more numerous. The conclusion then was that the emulsion must have interfered with the activity of the predator, possibly due to repellent odor (Craw 1894).

TWENTIETH-CENTURY DEVELOPMENTS

By 1900 the number of programs in biological control, the development of techniques for handling natural enemies, and the necessary facilities for such work had increased notably. Activity in California continued apace. In 1901 the parasite *Scutellista cyanea* Motschulsky, which a few years earlier had been introduced by the U.S. Department of Agriculture into Louisiana, was again imported into the United States, this time from South Africa to California for use against the black scale. In 1903 the California State Horticultural Commission constructed an insectary in San Francisco, a special facility designed to receive and propagate imported natural enemies. This was the first such specialized facility for supporting biological control work in the nation. Destroyed by fire after the San Francisco earthquake of 1906, the state insectary was relocated in Sacramento in 1907.

In 1904 the horticultural commission initiated a project for the biological control of the codling moth, importing the parasite *Ephialtes caudatus* (Ratzeburg) from Spain. The campaigns against the black scale and the California red scale, initiated soon after the Vedalia project, continued through the first decades of the new century.

Major projects were started elsewhere in the United States during this time. In cooperation with the state of Massachusetts, the U.S. Department of Agriculture began in 1905 a large-scale project against the gypsy moth, *Porthetria dispar* (Linnaeus). This serious lepidopterous pest of numerous deciduous trees and shrubs in the northeastern United States apparently invaded America about 1869. A noteworthy feature of this program was the construction of a gypsy moth parasite laboratory, designed to receive and process natural enemies of the pest as these were received from explorative work in Europe. At the same time a control program was started against the brown-tail moth, *Nygmia phaeorrhaea* (Donovan), in New England.

A pest of wheat in the midwestern United States, became another focus of attention. This was the greenbug, an aphid, *Schizaphis graminum* (Rondani). A native parasite, *Aphidius testaceipes* (Cresson), heavily attacks the pest in the southern parts of its range, but is much less effective in more northerly regions. University of Kansas scientists, in 1907, collected and distributed large numbers of dead parasitized aphids, known as mummies, into northerly fields in order to increase parasitism there. The results were variable, and claims of successful control of the greenbug as a consequence of this effort were confounded by other influences on the dynamics of the pest populations. A similar effort, aimed at collecting and distributing stocks of the native, aphid-feeding lady beetle *Hippodamia convergens* (Guerin) to various vegetable-growing areas of California, was begun in 1910 by the California State Horticultural Commission, with indifferent results, and was later found to be ineffective.

Elsewhere, biological control work increased from 1900 to 1910. As can be seen from Table 1, there were, throughout the world, at least eleven

Table 1. The progress of biological control work, 1890-1970, by decades. Each project judged completely or substantially successful by DeBach (1964) is tabulated at the time of the initial establishment of the natural enemy involved.

Decade	*Completely or substantially successful*	*Partially successful*	*Total*
1890-1900	1	1	2
1900-1910	7	4	11
1910-1920	6	8	14
1920-1930	17	11	28
1930-1940	32	25	57
1940-1950	10	12	22
1950-1960	9	5	14
1960-1970	14		14+

different programs of biological control, seven of which were judged to be completely or substantially successful. Countries involved include Australia, Hawaii, Italy, Peru (DeBach 1964).

QUARANTINE CONSIDERATIONS

It was soon realized that beneficial insect importations could be quite hazardous unless done with great care by experts using special precautions. The danger of accidental introduction of new insect pests, plant pathogens, or parasites of the natural enemies themselves during the routine handling of importation parcels suggested the need for strict quarantine security. The U.S. Department of Agriculture reserved to itself authority to permit shipments of insects and plants or plant parts from foreign sources into the United States. The first quarantine facility, designed specifically for this purpose, an insect-proof laboratory limited as to personnel access, was constructed in Hawaii in 1913. Within the next several decades all centers for the receipt of imported natural enemies were required by the U.S. Department of Agriculture to provide such facilities.

In California, in 1913, an experienced biological control specialist, Harry S. Smith, then employed at the gypsy moth parasite laboratory in Massachusetts, was appointed superintendent of the state insectary at Sacramento. Smith promptly increased foreign exploration activity on behalf of California's pest control efforts, and he himself set out soon after appointment on his first overseas mission, to Japan and the Philippines, in search of enemies of the black scale. In 1914, Smith employed H. L. Viereck to search in southern Europe for enemies of mealybugs. In Sicily Viereck discovered and sent to Sacramento by ship the small encyrtid parasite, *Leptomastidea abnormis* (Girault), a very effective enemy of the citrus mealybug. This parasite, first released in Southern California in 1914, soon spread widely, resulting in a renewed interest in the biological control method and the consequent intensification of work in the state insectary under Smith.

During the decade 1910-1920, more than a dozen cases of establishment of natural enemies for control of pests were recorded around the world. At least six of these were judged to have been completely or substantially successful (see Table 1), involving the following pests:

Brown-tail moth	Canada
Anomala beetle	Hawaii
Alfalfa weevil	Utah
Rhinoceros beetle	Mauritius
Sugar cane weevil	Hawaii
Larch sawfly	Canada

In 1919, the U.S. Department of Agriculture created a laboratory in France to serve as a base of operations for the collection, rearing, and identification of natural enemies of the European corn borer, a major pest of corn and other crops in central North America. To reduce the possibilities of accidental introduction of harmful organisms to America, corn borer natural enemies were reared to the adult stage and shipped free of their hosts, any of their own natural enemies (hyperparasites) which might be present, and all plant material which might harbor, undetected, other insect and disease pests. This meticulous care to prevent the transfer to the United States of noxious species during natural enemy importation work remains the hallmark of all biological control agencies.

As can be seen from Table 1, biological control work intensified through the years 1920 to 1940. In 1923 the biological control work of the state of California, still under the leadership of H. S. Smith, was transferred to the University of California Citrus Experiment Station, Riverside. A new quarantine-insectary facility was constructed there in 1929, and the old one in Sacramento was shut down. From this time on, as a consequence of Smith's enthusiastic and inspiring devotion to the study and promotion of biological control procedures, responsibility in California for the prosecution of all biological control work lay with entomologists in university service, an arrangement unique in the nation.

Smith's stimulating advocacy of biological control was not limited to California. His frequent contacts with pest control specialists in other nations inspired much additional biological control activity throughout the world. In the decade 1920-1930 more than thirty cases of natural enemy establishment were recorded. Among the important projects during this period was the dissemination of *Aphelinus mali* (Haldeman), the woolly apple aphid parasite, from its native home in New England, to New Zealand, where it soon provided very successful control of this apple pest, and then to British Columbia, Chile, South Africa, Italy, Uruguay, Brazil, Australia, and many other countries. Also noteworthy are the cases of control of the sugarcane leafhopper in Hawaii, the citrus blackfly in Cuba, the citrophilus mealybug in California, the greenhouse whitefly in Canada, among others. In the United States such major campaigns as the European corn borer project, the Japanese beetle program, and the oriental fruit moth project were started during this period.

The decade 1930-1940 saw the peak of activity throughout the world, with fifty-seven different natural enemies established at various places. At least thirty-two of these led to successful control results. Then, as can be seen in Table 1, there followed a sharp drop in biological control activity, mainly because of World War II. However, this drop was also a consequence of the substantial reduction in the United States Department of Agriculture program of biological control, presumably due to a feeling that returns were incommensurate with the efforts expended. The misguided, worldwide emphasis upon use of synthetic organic insecticides beginning in 1945 more or less preempted the revival of biological control on any significant scale.

On the other hand, in California, biological control work continued unabated. In 1945 a new biological control laboratory was established by the University of California at its Gill Tract facilities in Albany, near Berkeley. Initial focus of this new laboratory was on control of the oriental fruit moth and the weed St. Johnswort, or Klamath weed, as it was known in western North America. A quarantine facility was added at Albany in 1951.

DEVELOPMENT OF INTERNATIONAL ORGANIZATIONS

By the mid-1920s many of the dominions and colonies of the British Empire were active in biological control work, including Australia, New Zealand, Fiji, Canada, Bermuda, South Africa, and others. In some of these countries, biological control facilities and teams were permanently employed. At Belleville, Ontario, the Canada Department of Agriculture constructed a biological control laboratory in 1929, and in 1936 a quarantine reception center was added. However, in 1971 this facility was closed and the quarantine reception activity transferred to Ottawa.

In 1927, the Imperial Bureau of Entomology created a special facility for the conduct of biological control work in England, the Farnham House Laboratory. This institution, which in 1928 came under the direction of W. R. Thompson, accepted requests for study, exploration, and delivery of natural enemies for use against specified pests of importance to the various countries of the British Empire. Concentrating first on enemies of insect pests, in 1929 it broadened its scope to include insect enemies of weeds. At Farnham House Laboratory a variety of services were provided, including a bibliographic service pertaining to literature on biological control, the cataloging of important pests and their natural enemies, the preliminary surveying by field investigators of pest-infested areas for additional natural enemies and for ecological information about their distribution and abundance, the collection of living samples of these enemies and their identification, the determination of their life histories and requirements for culture, and the development of appropriate shipping techniques for transferring selected natural enemies to various recipient agencies throughout the world.

Early projects undertaken at Farnham House Laboratory included search for and study of enemies of the codling moth, *Laspeyresia pomonella*, the pear slug, *Caliroa cerasi* (Linnaeus), the European fruit lecanium, *Eulecanium coryli* (Linnaeus), the oriental fruit moth, *Grapholitha molesta* (Busck), the pine shoot moth, *Rhyacionia buoliana* Schiffermüller, sawflies of the genus *Sirex*, the larch case bearer, *Coleophora laricella* Hübner, the lucerne flea, *Smynthurus viridis* Linnaeus, the wheat-stem sawfly, *Cephus cinctus* Norton, the diamondback moth, *Plutella maculipennis* Curtis, the carrot rust fly, *Psila rosa* Linnaeus, the potato tuber moth, *Phthorimaea operculella* Zeller, the pink bollworm, *Pectinophora gossypiella* (Saunders), the greenhouse whitefly, *Trialeurodes vaporariorum* Westwood, the sheep blowfly, *Lucilia sericata* Meigen, and the horn fly, *Lyperosia irritans* Linnaeus. Stocks of the parasite *Aphelinus mali*, derived from the United States, were maintained for distribution of requestors for the control of the woolly apple aphid.

In 1940, the Empire biological control facility was moved because of the war to Ottawa, Canada, where it became known as the Imperial Parasite Service, still a component of the Imperial Institute of Entomology. In 1947 the service became independent, and was designated as the Commonwealth Bureau of Biological Control. A few years later, in 1951, the facility received its present name, Commonwealth Institute for Biological Control, or CIBC. In 1961, the CIBC headquarters were transferred to Trinidad, West Indies, where they remain today.

Starting with but one laboratory/office/insectary facility at Farnham Royal, United Kingdom, the CIBC now embraces not only an administrative headquarters and laboratory, but also regional stations and substations in Argentina, Switzerland, India, Malaysia, Pakistan, Uganda, West Africa and West Indies. With financial support from the Commonwealth countries, the institute carries out, by request, foreign exploration for beneficial species, studies on life histories and ecologies of various pests and their natural enemies in their areas of indigeneity, and the forwarding of shipments of enemies to requisitioners.

The CIBC was the first truly worldwide biological control organization. Besides providing services to Commonwealth countries, it also responds to requests from other nations at cost. The U.S. Department of Agriculture and the University of California have at various times received information and parasite materials from the CIBC and vice versa. In 1970 the program of the CIBC included research work centered at some 23 stations and substations, involving 118 different projects.

In 1948, under the auspices of the International Union of Biological Sciences, an international conference was held at Stockholm, Sweden, composed of biological control experts and administrators from many countries of the world, for purposes of considering the establishment of an international organization for biological control. A country, or individual, joining such an organization would be entitled to receive information and services about natural enemies of pest species, and would have the opportunity to acquire, at cost, shipments of beneficial organisms from other members or member organizations.

Because many countries and institutions in attendance at the meeting already provided for their own biological control services, including facilities and staffs of experts capable of supplying much of the expertise needed to carry out biological control programs, most, including the U.S. Department of Agriculture, California, Hawaii, the British Commonwealth, many Commonwealth countries, and Russia, decided not to join such an international organization. As a result, when in 1955 the Commission Internationale de Lutte Biologique contre les Ennemis des Cultures came into being, it included principally countries and members from European, Mediterranean, and Near Eastern regions. The CILB established headquarters at Zurich, sponsored a quarterly scientific journal, *Entomophaga* and maintained a taxonomic service in Geneva and a bibliographic service centered at Darmstadt, Germany. Working groups, composed of experts appointed by the general secretariat, met on occasion to consider and advise on possible solutions of particular problems, such as the biological control of the Colorado potato beetle in Europe.

In 1962, the CILB undertook a reorganization, changing its name to the Organisation Internationale de Lutte Biologique contre les Animaux et les

Plants Nuisibles. In 1971, again through the stimulus of the International Union of Biological Sciences, the OILB was further revamped into a true worldwide organization, composed of regional sections and a parent global organization. This organization has now been widely accepted as the definitive vehicle of biological control specialists on a worldwide basis.

REVIEW AND RESEARCH QUESTIONS

1. What historical concepts of an ecological nature preceded the development of practical biological control?

2. What were the earliest known cases of the intentional use by man of insects to control insect pests?

3. Contrast the development of biological control in Europe with its development in America. What was the major difference in pest origins in these two regions that led to the flowering of biological control in America?

4. What was the first international importation of a natural enemy to control a pest?

5. Discuss the origins of the quarantine reception laboratory in biological control work. Why is such a facility considered necessary?

6. Trace the development of the international biological control organizations.

BIBLIOGRAPHY

Literature cited

Clausen, C. P. 1940. *Entomophagous insects.* New York: McGraw-Hill, 688 pp.

Clausen, C. P. 1956. *Biological control of insect pests in the continental United States.* U.S. Dept. Agric. Tech. Bull. 1139, 151 pp.

Craw, A. 1894. *Biennial report of quarantine officer and entomologist.* Fourth Bienn. Rep. St. Bd. of Hort. St. of Calif. 1893-94, Sacramento.

DeBach, P., ed. 1964. *Biological control of insect pests and weeds.* London: Chapman & Hall, 844 pp.

Doutt, R. L. 1964. The historical development of biological control. In *Biological control of insect pests and weeds,* ed. P. DeBach, Chap. 2, pp. 21-42.

Essig, E. O. 1931. *A history of entomology.* New York: Macmillan, 1029 pp.

Gillet, F. 1882. *First report.* Calif. St. Horticult. Comm., Sacramento, pp. 24-28.

McCook, H. 1882. *Ants as beneficial insecticides.* Proc. Acad, Nat. Sci., Philadelphia.

Swan, L. A. 1964. *Beneficial insects.* New York: Harper & Row, 429 pp.

Sweetman, H. L. 1936. *The biological control of insects.* Ithaca, N.Y.: Comstock Publishing Associates, 461 pp.

Sweetman, H. L. 1958. *The principles of biological control.* Dubuque, Iowa: W. C. Brown, 560 pp.

4

Entomophagous Insects

The term *entomophagy* is a combination of the Greek words *entomon* (insect) and *phagein* (to eat), and thus denotes the insect-eating habit. There is also an adjectival form of this Greek word combination, as when we speak of *entomophagous* insects. Finally, the same Greek combination is used in a generic sense when we lump the insect-eaters together in the *entomophaga.*

In this volume our interest is largely with entomophagous insects. However, other kinds of animals also feed upon insects and often play important roles in insect population regulation. The Nematoda includes some important parasites of insects. Among the noninsectan arthropods there is a broad spectrum of insect-feeders, including scorpions, spiders, mites, sunspiders, pseudoscorpions, and harvestmen. In the higher groups there are entomophagous amphibians, fish, reptiles, birds, marsupials, and mammals, some of which are exclusively insectivorous. Even in man, the degree of entomophagy is surprisingly well developed (Bodenheimer 1951). Among the California Indians, oak wax scale, grasshoppers, wood-boring beetle larvae, June beetle adults, and crane fly larvae, fly puparia, adult flies, salmon fly nymphs and adults, a variety of caterpillars, and bee and yellowjacket larvae were all utilized as food (Essig 1931).

Insect-feeding in birds of various species has been well studied, and a number of species have been found to be of considerable importance in controlling forest insects (Buckner 1966, 1967, 1971). Shrews and mice are also important enemies of forest insects (Buckner 1966, 1967, 1971). Various fishes and amphibians are important in the biological control of aquatic insects, some of which (e.g., mosquitoes, gnats, midges) are pests (Gerberich 1946; Gerberich and Laird 1968; Sweetman 1958).

In applied biological control, several noninsectan species have been deliberately used with some measure of success. Included are *Gambusia* and other fishes against mosquitoes, gnats, and midges (Gerberich 1946; Gerberich and Laird 1968); the nematode *Neoaplectana glaseri* Steiner against the Japanese beetle; the giant toad *Bufo marinus* Linnaeus for control of white grubs and sugarcane rhinoceros beetle (Sweetman 1958); and the mynah bird, *Acridotheres tristis* (Linnaeus), against the red locust (Sweetman 1958).

But despite the wide spectrum of insect-eating organisms, the role of noninsectan species, particularly in classical biological control, has been minor, which is why the emphasis of this volume is on insects in biological control.

Entomophagous insects fall into two categories, *predators* and *parasites (parasitoids).* The two groups differ in several ways. Characteristically, a predator is relatively large compared to its host (prey), which it seizes and either devours or sucks dry of its body fluids rather quickly. Typically the individual predator consumes a number of hosts (e.g., a single lady beetle larva may consume hundreds of aphids) in completing its development. Most often a predator is carnivorous in both its immature and adult stages and feeds on the same kind of host in both stages.

By contrast, the parasitoid is almost invariably parasitic only in its immature stages, and develops within or upon a single host, which is slowly destroyed as the parasitic larva completes its development. The parasitoid adult is in most cases free living, feeding on such foodstuffs as nectar, honeydew, and sometimes host insect body fluids. For example, see Figure 5.

Figure 5. Parasitic flies—A, *Eucelatoria armigera* Coquillett; B, *Archytas californiae* (Walker). Larvae of these flies are parasitic in lepidopterous larvae, particularly certain cutworms and armyworms. Photographs by Ken Middleham, University of California, Riverside.

PREDATORS

Predatory insects are the lions, wolves, sharks, barracuda, hawks, and shrikes of the insect world. (See Figure 6.) The predatory insect either lies in wait to pounce upon its unsuspecting victim, runs it down, or where the host is sessile or semisessile it may literally browse off the population.

Predatory insects feed on all host stages: egg, larval (or nymphal), pupal, and adult. From the standpoint of feeding habit, there are two kinds of predators, those with chewing mouthparts (e.g., lady beetles [Coccinellidae] and ground beetles [Carabidae]), which simply chew up and bolt down their victims—legs, bristles, antennae and all—and those with piercing mouthparts, which suck the juices from their victims (e.g., assassin bugs [Re-

Figure 6. Typical predatory insects commonly encountered in biological control work. A, adult of the carabid beetle, *Calosoma affine* Chaudoir, feeding on larva of the western yellowstriped armyworm, *Spodoptera praefica* (Grote); B, adult of the damsel bug, *Nabis americoferus* Carayon, feeding on larva of the beet armyworm, *Spodoptera exigua*, on cotton; C, larva of the lady beetle, *Coccinella franciscana* Mulsant, feeding on a winged adult of the spotted alfalfa aphid, *Therioaphis trifolii*, on alfalfa; D, larva on the green lacewing, *Chrysopa carnea* Stephens, feeding on larva of S. *exigua* on cotton. Photography by F. E. Skinner, University of California, Berkeley.

duviidae], lacewing larvae [Chrysopidae], hover fly larvae [Syrphidae]). The sucking type of feeder often injects a powerful toxin which quickly immobilizes the prey so that the feeding process is a placid affair with little thrashing about by the victim. For example, once a lacewing larva clamps its sickle-like mandibles into a caterpillar several times its size, the latter is doomed and its period of struggle lasts but a few seconds.

Predatory species occur in most insect orders, with the greatest number of species occurring in the Coleoptera. One order, the Odonata (dragonflies) is exclusively predaceous. Predators may be *oligophagous,* having a broad host range (e.g., the green lacewing, *Chrysopa carnea* Stephens), *stenophagous* having a restricted host range (e.g., aphid-feeding coccinellids and syrphids), or *monophagous,* that is, highly prey specific (e.g., *Rodolia cardinalis,* which feeds only on *Icerya purchasi* and its close relatives). With *R. cardinalis,* the parasitic habit is approached, since the female beetle deposits its egg on an adult female scale or on a single scale's egg mass, and the hatching larva can complete its development on the eggs supplied by this initial source (Clausen 1940). A number of predatory insects resemble parasites in this way and there are a number of parasitoids whose habits approach those of predators. Many entomologists, in fact, consider the parasitoids to be specialized predators rather than parasites.

Although predators have been far overshadowed by parasites in classical biological control, they nevertheless have been of great significance in several programs, e.g., *R. cardinalis* versus cottony-cushion scale and *Tytthus (=Cyrtorrhinus) mundulus* (Breddin) versus sugarcane leafhopper, and their importance in naturally occurring biological control is inestimable. For example, in cotton in California's San Joaquin Valley, predators are much more widely important in restraining the major lepidopterous pests (i.e., bollworm, cabbage looper, beet armyworm) than are parasites. The worldwide eruption of spider mites in the wake of widespread use of chemical insecticide has mainly resulted from the elimination of predators of spider mites by the insecticides (McMurtry et al. 1970; Huffaker et al. 1970).

PARASITES

An insect which parasitizes other insects is known as a *parasitoid,* a term which distinguishes these entomophagous insects from all other kinds of parasites. No other group of organisms parasitizes its own kind to the extent that insects do. In fact, in insects this habit has reached the remarkable extreme of *adelphoparasitism*, wherein a species is parasitic on itself as with certain Aphelinidae in which males are obligate parasites of females of their own species (e.g., *Coccophagus scutellaris* Dalman, *Coccophagoides utilis* Doutt).

Parasitoids are recorded from five insect orders with the bulk of the species occurring in the Diptera and Hymenoptera. Despite their restriction to only five orders, there are tremendous numbers of parasitoid species worldwide. Kerrich (1960), extrapolating from the figures of described Coleoptera and parasitic Hymenoptera in the well-studied British fauna (4000 beetle species and 5000 species of parasitic Hymenoptera), speculated that there might be a half million described parasitic Hymenoptera worldwide if the global fauna had been as well studied as that of the beetles (of which, by his estimates, there were 300,000 described species in 1960).

Kerrich also stated that W. H. Ashmead's sixty-year-old guess of a million ichneumonids alone "still seems sensible today." Townes (1969) was considerably more conservative in his estimate of the number of ichneumonids, giving a figure of approximately 60,000 species, which is still a tremendous number. Although these numbers are only guesses, they are the projections of experienced scientists and even if they are an order or two in magnitude off the mark they reflect the enormous abundance and variety of the parasitoids.

Parasitoids attack and develop in all insect stages—egg, larval (nymphal), pupal, and adult—and again as with the predators the host ranges of individual species run the full spectrum from monophagy to polyphagy. (See Figure 7). For example, the tachinid *Compsilura concinnata* Meigen has been recorded from more than 100 hosts representing 3 orders and 18 families (Clausen 1940), while the aphidiid *Trioxys complanatus* Quilis is stenophagous, developing only in species of *Therioaphis,* and its close relative *T. pallidus* (Haliday) is apparently monophagous on *Chromaphis juglandicola* (Kaltenbach).

Kinds of parasitoids

Insect parasitism manifests itself in a number of ways, and as a result there is a considerable terminology which describes the kinds of parasites and the nature of their development.

Primary Parasite. Primary parasites are those species that develop in or upon nonparasitic hosts. These hosts may be phytophagous, saprophagous, coprophilous, polleniferous, fungiferous, predatory, etc., but in no case are they themselves parasitoids.

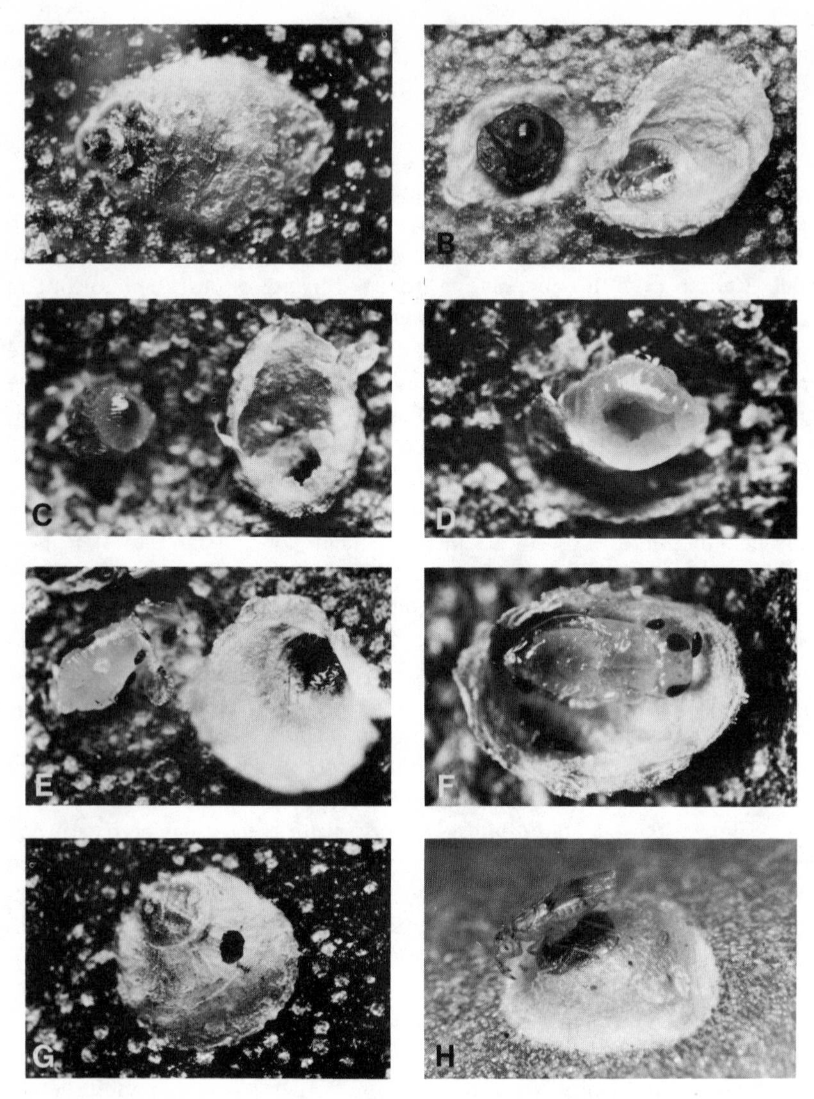

Figure 7. A typical parasitic insect associated with a scale pest. A, mature female olive scale, *Parlatoria oleae,* on upper surface of olive leaf; B, scale covering removed, showing young larva of the wasp, *Aphytis maculicornis,* feeding externally on the body of a mature female olive scale; C, older parasitic wasp larva, as in B; D, mature parasitic wasp larva, as in B; E, early pupa of parasitic wasp (showing black meconial pellets and empty skin of host, as in B; F, parasitic wasp pupa (showing eye coloration and meconial pellets), as in B; G, exit hole of adult *Aphytis* in female scale; H, adult *Aphytis* examining mature scale preparatory to oviposition. Photographs by F. E. Skinner, University of California, Berkeley.

Hyperparasite. A *hyperparasite* is a parasitoid which develops on another parasitoid (i.e., a parasite of a parasite). There may be more than one level of hyperparasitism in a given relationship. Thus, in the relationship between the pea aphid and *Aphidius smithi* Sharma and Subba Rao, the primary parasite *A. smithi* may first be attacked by *Alloxysta victrix* (Westwood), which for precision may be called a secondary parasite (Gutierrez and van den Bosch 1970), and *Alloxysta* in turn may be attacked by *Asaphes californicus* Girault, a tertiary parasite (Sullivan 1969). *Asaphes* may also directly attack *Aphidius* in which case it acts as a secondary parasite. Both *Alloxysta* and *Asaphes* are *direct hyperparasites,* since they oviposit directly into or upon the primary species. But the two differ in their method of attack. *Alloxysta* attacks its host (*A. smithi*) when the latter is a second-instar larva, still contained within the living aphid. In doing this, the *Alloxysta* female penetrates the aphid integument with its ovipositor, seeks out the *Aphidius* larva with this organ and deposits an egg internally in its victim. The hatching *Alloxysta* larva then develops as a solitary internal parasite of *Aphidius* and ultimately kills it in the prepupal stage. *Asaphes* attacks its host by drilling through the aphid mummy (parchment-like skin of the dead pea aphid), venomizing the contained *Aphidius* or *Alloxysta,* and then depositing an egg on the surface of the victim so that the hatching larva develops as an external parasite. As mentioned above, species such as *Asaphes* and *Alloxysta* are termed direct hyperparasites because of their direct oviposition into or upon the victim. By contrast, some hyperparasites simply oviposit into an insect whether it is parasitized or not. The hatching larva then seeks out the primary parasite if it is present, or if not waits until a primary parasite egg is deposited in the insect and attacks the hatching larva. If no primary parasite egg is deposited the hyperparasite simply perishes for lack of a host. Species with this type of habit are termed *indirect hyperparasites.* The rather complex relationships described are perhaps clarified by Figure 8.

Endoparasite. Parasitoids which develop within the host's body (internally) are called *endoparasites.* Where only a single larva completes its development in a given host, the species is termed a *solitary endoparasite.* Where several to many larvae develop to maturity in a single host, the term *gregarious endoparasite* is used.

Ectoparasite. A species which develops externally (the larva feeds by inserting its mouthparts through the victim's integument) is known as an *ectoparasite.* Again as with endoparasites there are solitary and gregarious ectoparasites.

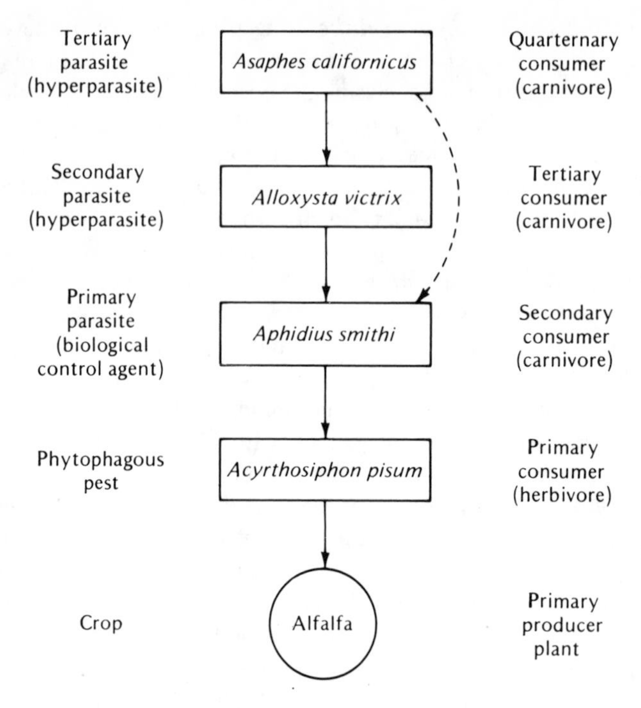

Figure 8. A plant, insect, insect parasitoid food chain of the type commonly encountered in biological control work, based on alfalfa, the pea aphid, *Acyrthosiphon pisum* (Harris), and its primary and secondary parasitoids. *Asaphes californica* may act as either a secondary or tertiary parasitoid as indicated.

Multiple parasite. A situation in which more than one parasitoid species occur simultaneously within or upon a single host is termed *multiple parasitism.* In most cases only one of these species survives to maturity, the others succumbing to competitive interaction (fighting, physiological suppression). In rare cases, as with species of *Trichogramma* which parasitize insect eggs (particularly those of Lepidoptera), more than one species may complete their development.

Superparasite. The phenomenon in which more individuals of a given parasitoid species occur in a host individual than can develop to maturity in that host is called *superparasitism.* Where this occurs with solitary endoparasites, internecine battle or physiological suppression of the supernumerary larvae or eggs results in the survival of a dominant individual. In some cases, however, the host itself succumbs prematurely before the supernumerary parasites are eliminated, and all perish.

Adelphoparasite (autoparasitism). The phenomenon in which a species of parasitoid is parasitic upon itself, is termed *adelphoparasitism.* This is the case with *Coccophagus scutellaris,* the male of which is an obligate parasite of the female.

Cleptoparasite. The phenomenon in which a parasitoid preferentially attacks hosts that are already parasitized by another species is called *cleptoparasitism.* The cleptoparasite is not hyperparasitic, for it does not parasitize the previously occurring parasite species. Instead, multiple parasitism is involved, and the relationship between the two species is competitive, with the cleptoparasite usually dominating.

Modes of reproduction

In the parasitic Hymenoptera several different patterns of reproduction occur which have important bearing on the ecology and habits of different species. These are all variations on a basic phenomenon in all the Hymenoptera known as *haploid parthenogenesis.* Haploid parthenogenesis refers to the fact that the unfertilized egg can undergo parthogenetic development to produce a normal, viable adult. In every case, the haploid individual is a male. On the other hand, the fertilized egg develops into a diploid female adult. However, there are differences in the way some parasitoids follow this basic pattern. Such differences in mode of reproduction are termed *arrhenotoky, deuterotoky,* and *thelyotoky.*

Arrhenotoky. The basic reproductive mode is *arrhenotoky,* in which unfertilized eggs produce males, and fertilized eggs produce females. Hence, virgin females can produce progeny but they will be all male. Species which follow this mode of reproduction are referred to, for obvious reasons, as *biparental.* It is important to note that in some biparental species the mated female can produce either male or female offspring through external or internal control of fertilization. In other species the mated female produces only female progeny.

Deuterotoky. The reproductive mode by which unmated females produce both male and female progeny is termed *deuterotoky.* The species thus produced are called *uniparental.* The haploid males produced in such cases are biologically and ecologically nonfunctional. The females retain the diploid condition through various cytogenetical mechanisms. Often in deuterotokous

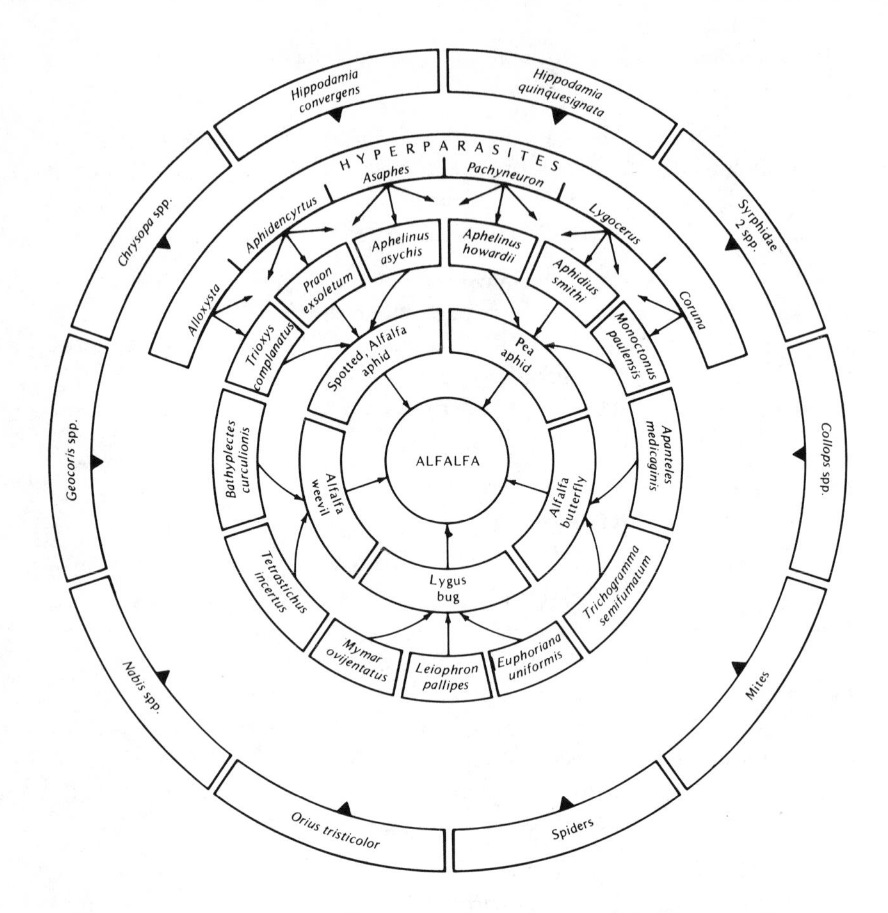

Figure 9. Entomophagous arthropods associated with certain insect pests of alfalfa in California. The insects listed in the ring surrounding the alfalfa circle are phytophagous pests. The next ring contains the main primary parasitoids of these pests. All parasitoids associated with the alfalfa weevil, spotted alfalfa aphid, and the pea aphid are introduced species; the others are native. The third ring includes hyperparasitic insects associated only with aphid-attacking parasitoids. The outer ring includes the predators, some of which (*Hippodamia* spp. for example) attack only aphids, including parasitized aphids, while others, general predators, attack most small-bodied inhabitants of the alfalfa plant.

species the proportion of males produced relative to total progeny varies with external conditions, such as temperature.

Thelyotoky. In *thelyotoky,* virgin females produce only female progeny, and males are unknown. Such uniparental species are not uncommon. Similar cytogenetical mechanisms for restoring the diploid condition in the egg occur here as in the case of deuterotoky. In a few cases, thelyotokous species, under stressful conditions of extreme temperature, for example, shift to the deuterotokous mode and produce haploid males as well as diploid females.

CONCLUSION

It is apparent from the foregoing discussion that entomophagous insects are varied and abundant, and that they frequently develop complex interrelationships with their hosts and among themselves. Figure 9 depicts the intricate relationship of several insect pests and their natural enemies in California alfalfa. As incomplete as it is, the diagram strikingly illustrates the wide spectrum of entomophagous arthropods that can impinge upon single phytophagous species in a simple agroecosystem. If we extrapolate from this model to the estimated ±10,000 pest insect species worldwide and consider the vast numbers of potentially injurious ones permanently restrained by natural enemies, the phenomenon of entomophagy assumes enormous significance both as to the diversity of predatory and parasitic species involved and the impact that they have on insects and insect-like arthropods.

REVIEW AND RESEARCH QUESTIONS

1. Define entomophagy and list some animals that exhibit this food specialization.

2. Contrast the insect predator with the insect parasitoid, and name several orders of insects that contain examples of each.

3. What is meant by the terms (a) *hyperparasitism* and *superparasitism,* (b) *endoparasitism* and *ectoparasitism,* and (c) *adelphoparasitism* and *cleptoparasitism.*

4. Which do you think would be the better performer, the arrhenotokous parasitoid or the thelyotokous parasitoid?

BIBLIOGRAPHY

Literature cited

Bodenheimer, F.S. 1951. *Insects as human food.* The Hague: W. Junk, 352 pp.

Buckner, C. H. 1966. The role of vertebrate predators in biological control of forest insects. *Ann. Rev. Ent.* 11: 449-470.

Buckner, C. H. 1967. Avian and mammalian predators of forest insects. *Entomophaga* 12: 491-501.

Buckner, C. H. 1971. Vertebrate predators. In *Toward integrated control,* pp. 21-31. Proc. Third Ann. N.E. For. Ins. Work Conf., USDA For. Serv. Res. Paper NE-194, 129 pp.

Clausen, C. P. 1940. *Entomophagous insects.* New York: McGraw-Hill, 688 pp.

Essig, E. O. 1931. A history of entomology. In *California Indians in relation to entomology,* Chap. 2, pp. 12-47. New York: Macmillan, 1029 pp.

Gerberich, J. B. 1946. An annotated bibliography of papers relating to the control of mosquitoes by the use of fish. *Amer. Mid. Nat.* 36: 87-131.

Gerberich, J. B., and M. Laird. 1968. *Bibliography of papers relating to the control of mosquitoes by the use of fish.* FAO Fisheries Technical Paper No. 75 (FR_S/T75), 70 pp.

Gutierrez, A. P., and R. van den Bosch. 1970. Studies on host selection and host specificity in the aphid hyperparasite *Charips victrix* (Hymenoptera: Cynipidae). 1. Review of hyperparasitism and the field ecology of *Charips victrix. Ann. Entomol. Soc. Amer.* 63: 1345-1354.

Huffaker, C. B., M. van de Vrie, and J. A. McMurtry. 1970. Ecology of tetranychid mites and their natural enemies: A review. II. Tetranychid populations and their possible control by predators: An evaluation. *Hilgardia* 40: 391-458.

Kerrich, G. J. 1960. The state of our knowledge of the systematics of the Hymenoptera parasitica. *Trans. Soc. British Ent.* 14(1): 1-18.

McMurtry, J. A., C. B. Huffaker, and M. van de Vrie. 1970. Ecology of tetranychid mites and their natural enemies: A review. I. Tetranychid enemies. Their biological characters and the impact of spray practices. *Hilgardia* 40: 331-390.

Sullivan, D. J. 1969. A study of aphid hyperparasitism with special reference to *Asaphes californicus* Girault (Hymenoptera: Pteromalidae). Ph.D. dissertation, on file University of California, Berkeley, 120 pp.

Sweetman, H. L. 1958. *The principles of biological control.* Dubuque, Iowa: Wm. C. Brown, pp. 369-393.

Townes, H. 1969. *The genera of Ichneumonidae.* Part 1. Mem. Amer. Entomol. Inst. No. 11, 300 pp.

Additional references

Askew, R. R. 1971. *Parasitic insects.* New York: American Elsevier, 316 pp.

DeBach, P. ed. 1964. *Biological control of insect pests and weeds.* London: Chapman & Hall, 844 pp.

Doutt, R. L. 1959. The biology of parasitic Hymenoptera. *Ann. Rev. Entomol.* 4: 161-182.

Dutky, S. R. 1959. Insect microbiology. *Adv. in Appl. Microbiology* 1: 175-200.

Nakagawa, P. Y., and J. Ikeda. 1969. *Biological control of mosquitoes with larvivorous fish in Hawaii.* World Health Organization WHO/VBC/69.173.

Sellier, R. 1959. *Les insectes utiles. Utilization des insects auxiliares. Utilization des insectas par l'homine.* Paris: Payot, 286 pp.

Procedures in Enemy Introduction

5

Classical biological control was described in an earlier chapter as the control of pest species by introduced natural enemies. The pattern to this process begins with determining whether the target pest is a native or exotic species and then passes through a series of steps involving foreign exploration, quarantine processing of collected material, mass propagation of the natural enemies, their field colonization, and finally the evaluation of their impact on the pest population. (See Figure 10.)

This chapter treats of the natural-enemy introduction process in considerable detail, for it is the very essence of classical biological control.

IDENTIFICATION OF THE PEST AS AN EXOTIC SPECIES

In the overwhelming majority of cases the target pests are exotic species. This circumstance evokes questions: How is a pest recognized as being exotic? How is its native habitat identified? What are the mechanics of exploration once the native habitat has been determined? These questions will be addressed and answered in the following pages.

In any contemplated natural-enemy introduction program, it should first be determined whether or not the target pest *is* an exotic invader. There is a dual reason for this: first, if the pest is indeed exotic the chances for its successful biological control are quite good, but if it is native there is little chance that it can be controlled by exotic natural enemies; second, once the pest is determined to be exotic, this knowledge is the key clue to its native habitat, the logical focus of search for its adapted natural enemies.

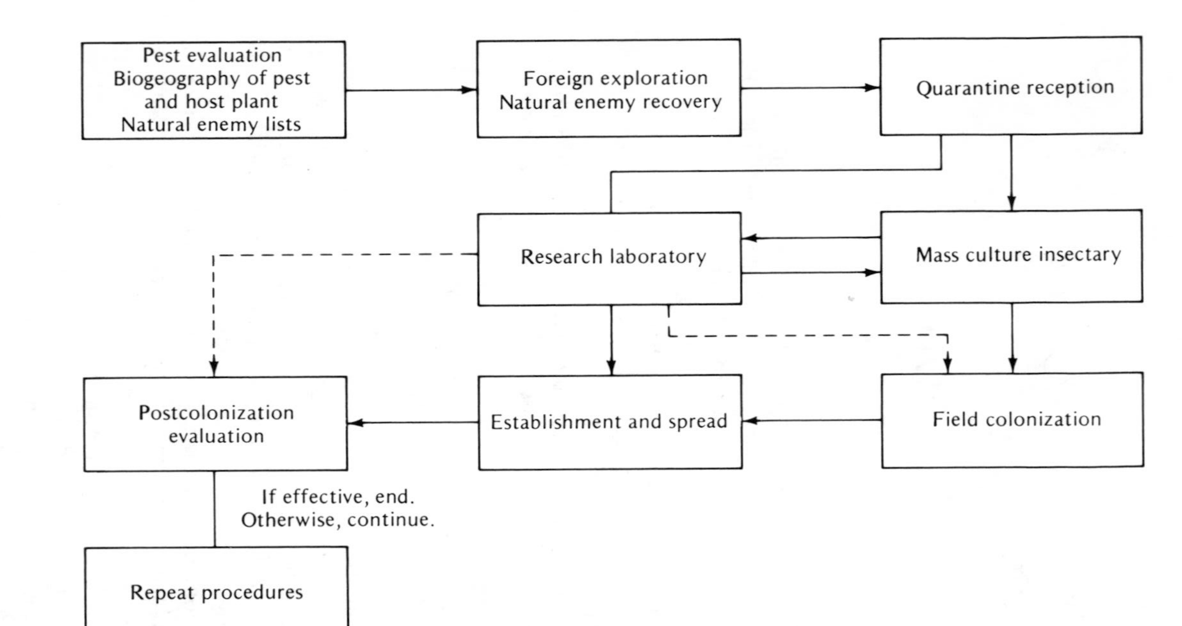

Figure 10. Block diagram describing the sequence of procedures and activities followed in a biological control campaign.

A number of indicators serve to identify the exotic invader, with the clearest signal being the abrupt outbreak of a species previously unknown in an area. This is not an infallible criterion, since some change in agricultural practice or the introduction of a new crop plant or variety could favor the eruption of a previously obscure indigenous species. But such an eruption is an infrequent occurrence, and a suddenly epidemic native species can usually be recognized, either because it has been previously known in the local fauna or because it belongs to an endemic taxon.

The really challenging cases of identification involve exotic species which have long resided in an invaded area or ones which have become cosmopolitan through accidental transfer by man. For example, most of the pest aphids in the United States are exotic species, but many of them have occurred in our crops for such a long time that they are literally accepted by farmer and entomologist alike as native species. Furthermore, many are so widely distributed globally that they are considered cosmopolitan and thus presumably Holarctic, when in fact most are really of Palaearctic origin.

How then do we recognize the exotic species from among these "old acquaintances" in the pest fauna? For one thing, a pest should be suspect when it is the sole representative of a taxon whose other members all occur in some remote land. Another strong indication of exotic orgin is a pest that lacks important parasites but is heavily attacked in another country; for example, the pea aphid *Acyrthosiphon pisum* (Harris) and the walnut aphid *Chromaphis juglandicola* existed for decades in California without important parasites. Heavy parasitization of these aphids by host specific parasites in the Palaearctic region provided strong evidence that they were native to the Old World. Finally, the host plant affinities of a pest can be an indication that it is an exotic species; for example, in North America the walnut aphid mentioned above is monophagous on *Juglans regia* Linnaeus (the so-called English or Persian walnut), an Old World tree, and never occurs on native *Juglans* spp., even though the latter often abound in the vicinity of commercial *J. regia* groves. This is strong evidence that the aphid originated in the same part of the Old World as did its host tree.

DETERMINATION OF THE NATIVE HABITAT OF THE EXOTIC PEST

Confirmation that a pest is an exotic species is only one part of the puzzle confronting the biological control specialist. The determination of the pest's area of indigeneity is equally critical for it is there that its full complement of adapted natural enemies occur. The case of the citrus black scale, *Saissetia oleae* (Bernard), serves to illustrate this point. This pest was long known to be an invader of California, but for many years its precise native habitat was unknown, since it had a worldwide distribution in tropical and subtropical areas. Thus, an array of ineffective natural enemies was imported into California from widely scattered areas including Brazil and Australia, before the outstanding parasite *Metaphycus helvolus* (Compere) was obtained from South Africa, the true home of the pest (van den Bosch 1968).

In most cases, the criteria used to identify a pest as an exotic species will reveal its area of indigenity. Taxonomic and other technical publications, museum material, the pest's host plant affinities, and the opinions of entomologists working with the species will usually, in the aggregate, reveal the true homeland. Our increasing knowledge of insect taxonomy and distribution makes it progressively easier for us to identify the homelands of invading species. But still there are some which are difficult to assign to specific areas of origin. The pink bollworm of cotton, *Pectinophora gossypiella,* is a case in point. With this insect, which occurs in most places on earth where cotton is cultivated, long accepted opinion holds that India is its native home. But the existence of a complex of *Pectinophora* species in Australia has caused some entomologists to question whether this is true. The matter is being investigated, but it is quite apparent that if Australia is indeed the area to which *P. gossypiella* is indigenous, much more promising parasites may be found there than in India.

Some pest species have such vast ranges over contiguous land areas that it is virtually impossible to pinpoint their original habitats. For example, the alfalfa weevil, *Hypera postica* (Gyllenhal), occurs over an estimated 8 to 10 million square miles in Europe, Africa, the Middle East, and Asia. This evokes the question: Does this represent the true area of indigeneity of *H. postica* or has man over the centuries accidentally spread it from a much more restricted focus?

The codling moth; *Laspeyresia pomonella,* the notorious "worm in the apple," is another species whose area of indigeneity in the Palaearctic region was originally almost surely more restricted than its distributional range there today.

It is very important for us to identify the original centers of distribution of such pests as the alfalfa weevil and codling moth, because it is there that we have the best chance of finding them in association with their full complexes of natural enemies. And with such pests this becomes something of a guessing game. Thus we might guess that ancient Media (northwestern Iran), the original home of alfalfa, may also be the original home of the weevil. But this is a somewhat shaky premise because the weevil attacks other species of *Medicago* and such clovers as *Melilotus* spp., *Trifolium* spp., *Trigonella* spp., which have different areas of origin. As for the codling moth, we should perhaps look to the original home of apple, which is presumably somewhere in Central Asia. But here again there is no guarantee that this will be the correct area since codling moth also attacks other fruits, e.g., apricot, pear, quince.

Whatever the case, it is of utmost importance that we utilize a variety of criteria in our attempts to determine the native habitat of a given pest. Reliance on a single criterion can lead to serious error. For example, when a highly destructive small yellow aphid invaded alfalfa fields in the southwest United States in the middle 1950s, it was identified as *Therioaphis maculata* (Buckton) and designated as the spotted alfalfa aphid. Buckton described *T. maculata* from material collected in India, and if the type of locality of the specimens from which he drew his description of this aphid had been the sole criterion upon which the search for its natural enemies was based, the effort would have met with failure, for as it turned out no parasites of the aphid were found in India. In fact, *Therioaphis maculata* is a synonym of *T. trifolii* (Monell), a species with wide distribution in Europe, Africa, and the Middle East (Hille Ris Lambers and van den Bosch 1964). It is now quite apparent that this aphid accidentally invaded India and that in doing so it escaped its effective parasites just as it did in invading the southwest United States. Fortunately, multiple criteria (e.g., advice and opinions of entomologists working with *Therioaphis* spp., information on parasites of the group, examination of museum material) were used in establishing the exploration program for parasites. As a result, collectors were sent to Europe, the Middle East, and Africa, as well as to India. Three species of parasitic Hymenoptera were obtained from Europe and the Middle East, colonized in the southwest United States, and ultimately came to play important roles in the biological control of the aphid (van den Bosch et al. 1964).

IMPORTATION AGENCIES IN THE UNITED STATES

In the United States four agencies are intensively involved in the exploration for and importation of natural enemies. These are: the Insect Identification and Parasite Introduction Branch of the Entomological Research Service of the U.S. Department of Agriculture, the University of California, the Hawaii Department of Agriculture, and the Hawaiian Sugar Planters Association. Each follows a somewhat different modus operandi but all have the same objective, the control of insect pests and weeds by imported natural enemies. Basically, each agency places explorers in the field in foreign areas, and these men then collect promising parasites, predators, pathogens, and weed-feeding species and transship them to quarantine laboratories where they are tested for desirable biological characteristics before being released to insectaries for mass production and ultimate colonization in the field.

In the United States all material shipped to the respective quarantine laboratories is by special permit issued by the Plant Quarantine Division of the U.S. Department of Agriculture. The University of California, the Hawaii Department of Agriculture, and the Hawaiian Sugar Planters Association operate under memoranda of agreement with the U.S. Department of Agriculture which stipulate, among other things, that certain safeguards are to be followed, adequate quarantine facilities provided, and qualified personnel employed in both the collection and quarantine processing of the imported material.

Material dispatched from overseas areas may be transported by several types of carrier, e.g., international airmail, diplomatic pouch, air express. But in all cases, the material is specially packaged so as to minimize possible escape of living insects or loss of extraneous plant matter from the parcels.

The efficacy of this "fail safe" shipping system is evidenced by the fact that though there has been an almost continuous movement of living insect material into the United States by one or another or all of the biological control agencies since the late 1880s, there has never been an accident in transit permitting escape and establishment of an undesirable species.

As indicated above, the U.S. Department of Agriculture, the University of California, the Hawaii Department of Agriculture, and the Hawaiian Sugar Planters Association each has a somewhat different modus operandi in the foreign areas. The U.S. Department of Agriculture personnel primarily operate out of two permanent laboratories in Europe, one near Paris that specializes in the collection and processing of entomophagous insects, and the other in Rome that is concerned with insects affecting weeds. At times U.S. Department of Agriculture personnel are temporarily posted to areas outside Paris or Rome for periods ranging from several weeks up to several years, but most activity is based at the two permanent laboratories.

University of California personnel are permanently stationed in the Divisions of Biological Control at Riverside and Berkeley (Albany), where they are responsible for various projects involving crop pests, forest insects, pests of man and animals, and weeds. Literally all foreign exploration is done by the responsible project leader or project personnel, and all such assignments are of limited duration (e.g., from several weeks to about one year). These persons usually hold collaboratorships in the U.S. Department of Agriculture and thus (through mutual arrangement) may work out of one of the U.S. Department of Agriculture overseas facilities. At other times, they may make arrangements with foreign institutions (e.g., the Commonwealth Institute of Biological Control, or various universities or ministries of agriculture), or they may carry out their activities completely independently.

In the University of California system the natural-enemy procurement, propagation, colonization, and evaluation sequence is a totally integrated process under the supervision of the responsibile project leader. This is possible because even though California is a large state, its croplands, forests, and other insect-affected areas are readily accessible to project personnel. Thus, once a natural enemy has cleared the quarantine laboratory, its mass propagation, colonization, and eventual evaluation can be easily coordinated by a single person. This gives an essential unity to the whole program.

In the U.S. Department of Agriculture, the entomophagous insects are most often processed through quarantine in Moorestown, New Jersey, and then usually sent to a laboratory located in that part of the country where the pest problem occurs to be mass-propagated there. Even then, material to be released may again be shipped hundreds of miles across state lines to substations or interested researchers if the pest is one that is of regional distribution (e.g., over the corn, wheat, or cotton belts). Such a scattered operation makes it difficult for a single person to coordinate a program. Thus, once the material has been processed through quarantine, the responsibility for mass propagation, colonization, and evaluation usually rests with one of various other research branches.

On the other hand, the U.S. Department of Agriculture biological control of weeds program is a highly integrated one. The personnel in the Rome laboratory (and a temporary one in Argentina) collect the candidate insects and conduct studies on their biologies and host plant affinities. Then when a species has satisfied the basic criteria, it is shipped to the U.S. Department of Agriculture quarantine laboratory at Albany, California, where further testing is conducted. When an insect is cleared for field release, its mass propagation, colonization, and evaluation are carried out by project personnel at Albany or by persons collaborating with these researchers.

The Hawaii Department of Agriculture operation resembles that of the University of California excepting that foreign exploration has traditionally been the responsibility of a single individual, a roving, permanent collector, who is literally on the road full time. Quarantine screening, mass propagation, and colonization are carried out by the Hawaii Department of Agriculture, and the evaluation studies are shared with personnel of the University of Hawaii. The Hawaii Sugar Planters Association operation almost exactly duplicates that of the University of California.

FOREIGN EXPLORATION FOR NATURAL ENEMIES

The search for and procurement of natural enemies is simply specialized insect collecting. The person assigned to a program is invariably a trained entomologist who may or may not have had previous experience in foreign collecting. But typically by the time the explorer undertakes a given assignment he will have developed a broad knowledge of the target pest and related species, as well as their known or possible natural enemies. He will be able to recognize the pest in all of its life stages, and he will possess as thorough an understanding of the pest's biology, ecology, phenology, and ethology as possible. In other words, when the explorer moves into the field, he knows what he is looking for and has an excellent idea of where to find it. This is not to imply that the search for natural enemies is an easy task. Indeed it is not. Often the pest is rare, or it is one of a complex of species and must be distinguished from its congeners. Thus, the discovery and recognition of the pest and its parasites requires knowledge, determination, and stamina, particularly if the insects are rare or if they occur in remote or inaccessible places.

Depending on the pest species involved, and the types of natural enemies being sought, collecting may be done in a variety of environments, e.g., parks, arboretums, streetside vegetation, home gardens, weedy places, orchards, pastures and rangelands, row crops, or undisturbed native vegetation.

The collecting techniques embrace the full spectrum of methods familiar to the trained entomologist: sweeping, beating, rearing (from infested fruits, leaves, stems, or seeds), trapping, handpicking of host insects and/or their parasitized or diseased stages and the various stages of the predators and parasites themselves.

Generally, parasite material is shipped in an inactive stage or in a minimally active one (e.g., as pupae, diapausing larvae, developing larvae within parasitized hosts), but sometimes it is necessary to send adult forms. Here moisture and nutrients must often be provided in the shipping container. Predators are also sent in their various developmental stages. Preferably, pupae or other resting stages should be sent, but often it is necessary to send active adults, nymphs, and/or larvae. Here again, nutritional and moisture requirements must be considered.

The duration of a foreign exploration program and the area covered will vary with the nature of the problem. Where a search is being undertaken for the first time, it should be programmed to extend over at least one full season's activity of the host species and to cover as much of the pest's distributional range as possible.

With some species this cannot be done in a single season or year and the natural enemy search may necessarily extend over an indefinite period of time. This is particularly true in weed control programs or where a large complex of natural enemies is involved and there appear to be important intraspecific variants of given species. On the other hand, where a particular natural enemy is being sought and its phenology is understood, the foreign collecting may extend over a period of only a few weeks or even a few days.

QUARANTINE RECEPTION

The routine introduction of natural enemies from foreign countries for purposes of biological control is fraught with potential hazards. For one thing it frequently happens, mostly unintentionally, but occasionally by necessity, that a noxious species, such as a foreign pest, or a phytophagous species not hitherto recognized as a pest, or an undetected plant pathogen carried in host plant material or in soil, may arrive in the same shipping parcel with the desired natural enemy. In addition, it is possible for such shipments to contain hyperparasitic species, or species which attack not the phytophagous pests of crops but instead the imported primary parasites or predators or natural enemies which already occur in the target region.

The escape of any of these unwanted species could well result in the establishment of a new pest in the target area, or of a hyperparasite which may interfere with the effective biological control by some beneficial species already established or the one to be colonized.

Therefore, it is customary, as in the United States, for the reception of shipments of beneficial organisms from overseas to be allowed only under permit, and then only when assigned to an authorized quarantine laboratory facility. In the United States, these quarantine laboratories are few in number, three operated by the U.S. Department of Agriculture at Moorestown, New Jersey, and Columbia, Missouri (natural enemies of pest insects), and Albany, California (natural enemies of weeds), three operated by the University of California at Albany, Riverside, and Parlier, California, one operated by Virginia Polytechnic Institute at Blacksburg, Virginia (for natural enemies of weeds), and one operated by the Hawaii State Department of Agriculture at Honolulu. (See Figure 11.)

To protect against accidental introductions of noxious species, a number of activities are carried on in the quarantine laboratory before a desired beneficial species is passed out for further propagation and/or colonization. These activities include the rearing and separation by species of adult parasites and predators from the imported sample, the careful taxonomic identification of the one or more species so recovered, the destruction of any remaining living host specimens and plant material, the culture of recovered beneficial species on locally derived host stocks, the preliminary working out of the cultural requirements and life cycle characteristics of the beneficial

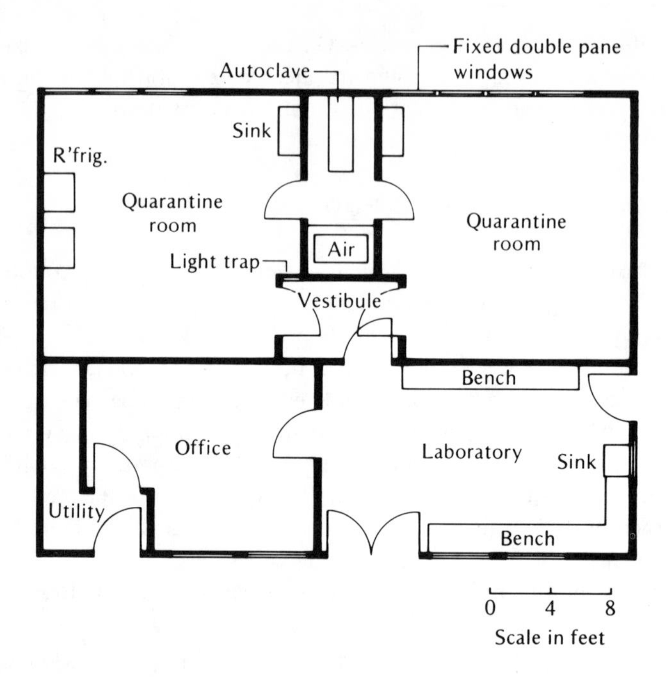

Figure 11. Floor plan of Quarantine Laboratory at the Division of Biological Control, University of California, Albany.

species, including observations of the reproductive habits, host selection, sex ratios, physical requirements, adult food needs, and so on. Only when all individuals of the species have been identified as to genus, transferred to a local host, and reared in pure culture (thereby ensuring against the obligate hyperparasite) are they allowed to be removed from quarantine. Customarily such pure cultures are transferred to the insectary for mass propagation and eventual colonization. Occasionally the beneficial species is passed on to a research laboratory for further study of adaptability to local conditions or hosts.

Because of the stringent security required in their operation, quarantine laboratories are restricted as to access. Only authorized quarantine specialists are allowed entry.

The work of the quarantine specialist is of fundamental importance, for it is upon his activities and capabilities that the success of the foreign explorer and the whole program depends. In some cases the foreign explorer is only able to ship a very few individuals of a given entomophagous species with unknown habits and needs. Starting with just these few specimens, the quarantine specialist must be able to carry out all of his examinations, handling techniques, cultural experiments, and life history observations. A technical failure involving the loss of a species is costly, not only in time and funds expended, but also from the standpoint of the potential benefit that the species might have brought.

MASS CULTURE OF ENTOMOPHAGOUS INSECTS

On occasion, important exotic natural enemies have become established in new areas after only very limited colonizations. For example, the famous *Rodolia cardinalis* became established following the release of but a few hundred individuals. Other examples of this sort are *Bathyplectes anurus* (Thomson) on the alfalfa weevil in the eastern United States, and the Iranian strain of *Trioxys pallidus* on the walnut aphid in California.

But, despite these "instant" successes, in most of the cases where species have become established, heavy and repeated colonizations were involved. Indeed, the mass production of introduced species is a crucial aspect of any classical biological control program, because it permits the release of sufficient numbers of the species at particular colonization sites, provides material for colonization in a variety of environments over a region (i.e., multiple colonizations in space), and allows for repeated colonizations over time.

Another aspect of natural-enemy mass culture involves production of material for inundative (overwhelming) or innoculative (reestablishment) colonizations in sustained programs, such as those practiced in citrus groves in parts of southern California. In these programs such species as *Cryptolaemus montrouzieri,* a predator of the citrus mealybug, *Metaphycus helvolus,* a parasite of the citrus black scale, and *Aphytis melinus* DeBach, a parasite of the California red scale, are maintained on a permanent basis in the insectary and mass cultured as needed for releases in groves where their hosts have risen or threaten to rise to injurious levels.

For the several reasons outlined, those organizations which are engaged in biological control maintain insectaries and supporting facilities and staff for the purpose of mass culture of natural enemies. (See Figure 12.)

The culture of entomophagous insects and their hosts is a complex

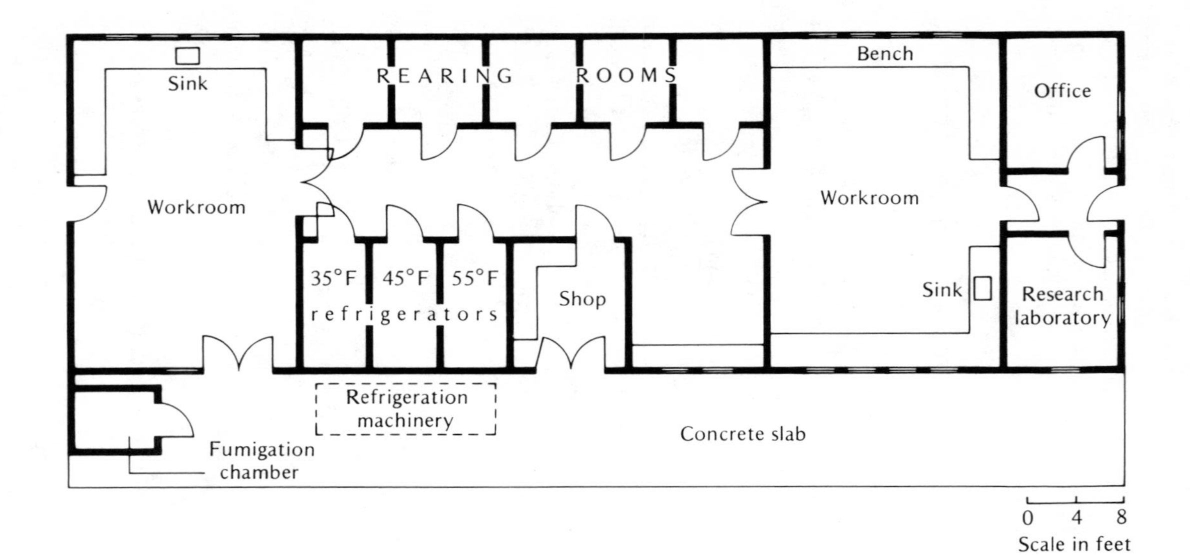

Figure 12. Floor plan of a typical mass-culture insectary, modified after that located at the Division of Biological Control, University of California, Albany.

matter, and the techniques and manipulations employed necessarily vary with each species being propagated. In the mass propagation process, there are two major concerns: (1) the provision of abundant host material for culturing given natural enemies, and (2) the development of techniques to assure maximum reproductive activity and optimum development and survival of the natural enemies. In other words, techniques must first be developed to propagate the insectary host in sustained volume, and then attain efficient propagation of the natural enemy.

Each program poses unique problems which must be worked out. In some cases easy solutions are reached, while in others great ingenuity and perseverance are required before satisfactory propagation is attained. Thus, with parasites of certain diaspine scales, long experience has led to the development of a rather standardized rearing technique usually involving use of factitious hosts that are amenable to insectary culture. Two scale species that have been particularly important in this role are the latania scale, *Hemiberlesia lataniae* (Signoret), and the oleander scale, *Aspidiotus hederae* (Vallot). In California these scales, grown on potato tubers or banana squash, have proven to be excellent insectary hosts for parasites of the California red and San José scales. This has been of critical importance because these latter scale insects are difficult to mass propagate under insectary conditions and are not biologically optimum hosts in the artificial environment. Fortunately, the several parasites of the California red and San José scales accept factitious hosts in the insectary and develop on them. This has permitted efficient mass propagation of these parasites.

Unfortunately, most natural enemies cannot be propagated on other than their natural hosts, which means that techniques must often be developed to culture the pest's host plant, the pest itself, and finally the natural enemy or enemies. Needless to say, such triple-phased programs pose a great challenge to the insectary specialist. For example, when the decision was made to attempt biological control of the walnut aphid in California, the insectary team was confronted with three major problems. First, they had to develop techniques for maintaining, in good supply, foliated walnut seedlings the year around. Next, they had to develop a program of sustained aphid production. And finally, since at the outset it was unknown just what kinds of natural enemies would be imported or how many species, they had to be prepared to quickly develop rearing techniques for whatever was handed them from the quarantine laboratory. The main problem associated with host plant culture was that of maintaining a continuous supply of seedlings in foliage. Walnut is a deciduous tree and once the plants drop their leaves they will not regenerate foliage unless conditioned by cold. This was accomplished by holding the dormant seedlings at 38°F for about six weeks, after which they were returned to the greenhouse where they then produced new foliage.

A continuous supply of foliated seedlings was maintained by rearing the plants in staggered lots (they were grown from nuts planted in vermiculite in pots and watered with a nutrient solution) and rotating these staggered lots from greenhouse, through insectary, through refrigeration, back to the greenhouse. The major obstacle to continuous production of aphid and parasite was the potential for each to pass into diapause. This was prevented by maintaining the aphid culture under a twelve-hour photoperiod and the parasite culture under constant light. As a result of these manipulations and others of lesser general importance, a highly satisfactory mass propagation program was developed for the parasite *Trioxys pallidus,* several hundred thousands of which were released during the colonization program.

There are, of course, even more formidable problems in insectary propagation than those encountered in the walnut aphid program. For example, at Albany after several years of effort we still have not developed sustained production of certain parasites of the walnut husk fly and the alfalfa weevil. Here again, diapause is the key problem. Nevertheless, stocks of these parasites have been maintained and colonizations are being made; they simply aren't being made at maximum levels. The important point here is that although perseverance, ingenuity, and vigilance have all been applied to the several types of programs described, equal results have not been attained. But the formula works to full effect in most cases, and it is the only way to go about the business of mass propagation of natural enemies.

COLONIZATION

In any biological control program the entire foreign exploration effort, natural enemy transshipment, and quarantine processing and rearing activities may well go for naught if the candidate species are not properly colonized. Failure at this point is not only frustrating, but it is wasteful of time, energy, and finances. A number of factors militate against successful natural enemy introduction and are discussed in Chapter 6. Certain of these factors are beyond human control, and if they are sufficiently adverse they will simply preclude establishment of a given species no matter what degree of care is taken in colonization. But since at its beginning there is no way to know that a program faces impossible odds, any colonization effort must be carried out in such a manner that the natural enemy will have maximum opportunity to become established.

If there is a standard procedure to be applied to natural-enemy colonization it should involve the following: (1) the establishment of optimum conditions for natural-enemy performance at the given colonization sites, (2) use of adequate numbers of natural enemies in given releases, (3) a sequence of such releases at each site, and (4) the establishment of colonization sites distributed over the geographical or ecological ranges of the target pest.

With each of these factors there is really no practicable way to determine precisely just what is adequate or optimum. Without question, the most important determinant here is the sustained production of large numbers of natural enemies in the mass propagation insectary. Maximum production of natural enemies will quite obviously permit the colonization of "sufficient" numbers over a period of time at a large number of colonization sites.

In practice, the procedure most commonly followed is to select, in a variety of climatic situations, several typical sites where the pest insect occurs, to note the population dynamics and life history pattern of the pest at these sites, and then to release vigorous stocks of the natural enemy in the immediate vicinity of the appropriate stage or stages of the pest. The number of such sites to be selected and the rates at which the natural enemy is released at each site are determined by the availability of material from the insectary and the rapidity with which recoveries are made.

For example, in an ecological mosaic such as California, where we are currently introducing parasites of alfalfa weevils, there are four distinct climatic areas where intensive colonizations are being made: (1) the intensely hot and arid Colorado Desert Valleys, (2) the mild coastal area, (3) the hot and arid Great Central Valley, and (4) the transmontane (Great Basin) area of the state's northeastern corner. In each area colonization sites have been established in alfalfa on properties of cooperating farmers. Furthermore, in some of the areas additional sites have been established in uncultivated areas, principally in bur clover, (*Medicago hispida* Gaertner), a widespread wild host plant of the weevils. Wherever possible, an effort is being made to colonize substantial numbers of the several parasite species in a sequence of releases. The colonizations in alfalfa are being done in collaboration with Agricultural Extension Service personnel who arrange for release sites in commercial alfalfa fields in which the growers allow the colonized areas to remain unmowed and untreated with chemical insecticides through the first two croppings following colonization. Thereafter, normal mowing and control practices are resumed and the parasites must fend for themselves under these conditions.

The alfalfa weevil parasite colonization strategy will continue either until it has been determined that the parasites have become established or until it is felt that they do not have the ability to do so under the given conditions. In the latter event, more stringently controlled colonization sites may be established, as was done in the program against the spotted alfalfa aphid. In that program, so many hazards faced the released parasites in commercial alfalfa fields that after a year of intensive colonization efforts there was no evidence that any of the three introduced species had established a foothold at any of a large number of colonization sites. Among other things, the repeated mowing of the release fields, frequent chemical treatment of the plots by growers despite prior promises not to do so, and recurrent heavy attacks by coccinellids on the aphid populations (which caused violent fluctuations in aphid numbers) apparently precluded establishment of the parasites (van den Bosch et al. 1959). Consequently, arrangements were made with an alfalfa grower to establish a colonization site in one of his fields, over which complete control could be exercised so that conditions as near optimum as possible for parasite survival and reproduction could be maintained. The grower committed a 38-acre field to the program and gave assurances that under no circumstances would he apply insecticides to any portion of the field. He further agreed to set aside an area of approximately 5 acres as an experimental area in which any type of manipulation or cultural practice deemed necessary to favor establishment of the parasites could be employed. Within the 5-acre experimental area approximately 1 acre was established as the parasite release plot and therefore left uncut during the entire colonization period. The remaining portion of the 5-acre area was to be used as a parasite "nursery" if and when spread occurred from the 1-acre release point. This was to be accomplished through staggered cutting of the alfalfa which, it was felt, would favor the continuous occurrence of aphids in the area.

Perhaps the key factor leading to the ultimate establishment of the parasites in this plot was the placing of a large (6 by 6 by 3 feet) organdy-covered cage in the center of the plot into which the parasites were released. The cage was utilized in order to prevent coccinellid predation on a localized portion of the aphid population. To further favor the parasites, stocks of aphids were periodically introduced into the cage to maintain a consistently high infestation level on the confined alfalfa. Parasites were introduced into the cage at intervals over a three-month period (April through June, 1955). When abundant mummified aphids were observed on the enclosed plants, the cage was opened to permit spread of the emerging adult parasites from this focus.

In this case the great care taken in managing the controlled spotted alfalfa aphid colonization site and in manipulating the alfalfa in the surrounding area proved to be the key to the successful establishment of the three parasites. By the end of the summer the parasites had spread into and heavily populated the entire 38-acre field, and they had even moved beyond its boundaries to adjacent fields. In the early autumn, mowed alfalfa bearing millions of parasitized, mummified aphids was trucked to other areas of

California and scattered in aphid infested fields. In this manner, rapid and widespread establishment of the parasites was accomplished from the single source created at the controlled release plot (van den Bosch et al. 1959).

The spotted alfalfa aphid parasite colonization program represents a situation where extreme care and deliberate manipulation was necessary in order to effect establishment of imported natural enemies. The program involving colonization of the Iranian strain of *Trioxys pallidus* versus the walnut aphid illustrates the opposite extreme. In this case very small liberations of *T. pallidus* at a limited number of sites, mostly on streetside and backyard walnut trees, resulted in immediate establishment and rapid increase and spread of the wasp in the season of liberation, even though colonizations were made at a time of year when the aphid population was at its lowest ebb. Furthermore, in the following year, when large-scale colonization of insectary propagated material was planned in commercial walnuts, only a few small, scattered liberations were realized because of an unanticipated breakdown in the insectary program. Nevertheless, the wasp became established everywhere that it was liberated, and it increased explosively (van den Bosch et al. 1970). Within two years of its initial colonization *T. pallidus* had spread over virtually all of the walnut-growing areas of central and northern California and had attained very important status as a natural enemy of the aphid.

But despite such an easy success, the experience from a majority of the colonization programs suggests that establishment will ordinarily be most difficult to attain, and great care and effort must be expended in the attempt to achieve this goal.

EVALUATION OF NATURAL ENEMIES

The question is often asked, both by the biological control expert and the skeptic, whether any suppression of a host pest population subsequent to the establishment of a biological control agent can be attributed to that natural enemy. In the earlier days of biological control the simple reduction in pest abundance, that is, the "before and after" pattern of pest density, was commonly cited as the demonstration of natural-enemy efficacy. It was also once customary to use the increase in levels of parasitization or predation by the newly established enemy as a measure of effectiveness. However, it has since been shown that percentage parasitization or predation cannot be equated with level of control exerted because it is the number of survivors that escape natural-enemy action that determines subsequent pest density, and not percentage mortality. The awareness that some new pest invasions and outbreaks eventually subside to lower levels even when no biological control effort is made, coupled with the criticism by some that biological control rarely worked and that any pest reduction following natural-enemy importation was due more likely to changes in farm practices, in the crop plant, or in the properties of the pest, or to the occurrence and value of already established, native natural enemies, led to attempts to evaluate natural-enemy action by experimental and analytical means.

The evaluation of natural-enemy efficacy has followed two patterns. On the one hand there are experimental procedures, the enemy exclusion or "check" methods which are covered in detail below. The other evaluation approach is the analytical one which makes use of the life table technique. This will be explained later.

Natural-enemy exclusion methods

Four different exclusion methods have been used which can demonstrate the effectiveness of natural-enemy action in particular circumstances. Each has certain limitations or objectionable aspects. As will be seen, the method appropriate to the case must be determined by the nature of the crop, the pest, and the enemies involved.

The mechanical barrier method. In this method a cage, cloth sleeve, wire screen, or similar device is used to enclose an uninfested plant, branch, section of bark, or field plot, and the pest is introduced into the protected zone. An adjacent, infested plot, either not enclosed or enclosed similarly as the above but with openings or other alterations in construction enabling free access of natural enemies, is used as a comparison check. In these situations, the protected pest population almost invariably tends to increase in density, while that in the "unprotected" check plot usually remains at the previous customary low level. The difference in density is ascribed to the action of the excluded natural enemies (DeBach and Bartlett 1964).

This technique has been applied to black scale on citrus, by use of organdy cloth sleeve cages, in evaluations of the parasite *Metaphycus helvolus* in southern California; to balsam woolly aphid on fir bark, by use of wire screen barriers to evaluate aphid predators; to *Aphis fabae* Scopoli on bean plants, by use of plastic screen cages; and to potato aphid *Macrosiphum euphorbiae* (Thomas) on potato plants, to compare the effect of the presence and absence of coccinellid and syrphid predators on aphid-caused plant damage and tuber yield. The procedure has not been restricted to insects; e.g., wire guards have been used to protect barnacles from marine predation in tidal zones.

In certain cases the barrier exclusion technique has included treatment of the inside surfaces of the barriers to kill off natural enemies trapped inside a cage. The host pest is prevented from coming in contact with the lethal cage, usually by virtue of its sessile (scale or mealybug species) or sedentary (aphids, mites) habits.

Objections to the use of cloth or screen cages to protect host infestations from natural enemy attack are that the internal microclimate of the protected plot has been altered in favor of the host. So long as barriers tight enough to prevent access of the usually very small natural enemies are used, this objection can only be forestalled by making sure that the "check plot" has a similar microclimate. This is not always possible. The environment can almost always be expected to be altered such that the efficiency of the natural enemy in controlling the pest is changed.

Chemical exclusion method. In this method a selective pesticide is used to eliminate or greatly inhibit the natural enemy, leaving behind, largely unaffected, the host infestation. Growth of the latter in comparison with a host population in an untreated check plot provides evidence of natural enemy effectiveness (DeBach and Bartlett 1964).

This technique has been applied successfully to the long-tailed mealybug, the cottony-cushion scale, the California red scale, the yellow scale, and the six-spotted mite on citrus, the cyclamen mite on strawberry, spider mites on apple, olive scale on olive, and cabbage root fly on cabbage. It eliminates the objection concerning altered microclimates and cage interference with host movements. However, it has been alleged, and even demonstrated in a few cases, that the selective pesticide, for example, DDT in the case of spider mites, scales, mealybugs, and aphids, or its associated spray adjuvants, may very well directly stimulate reproduction and population growth of the host species which to this extent would diminish or mask any measure of effectiveness of the natural enemy. It has also been claimed that the pesticide may stimulate the host pest indirectly through a physiological effect on the host plant. This, of course, can be proven or disproven by the appropriate experimentation. A version of this technique, designed to eliminate these objections, makes use of a pesticide barrier or border, at the periphery of the study plot. Natural enemies inside the barrier eventually move out to it and are destroyed; enemies outside the barrier zones are prevented from entering the study area. This technique works best with sessile hosts and mobile enemies.

Biological check method. This method, applicable only in certain limited situations, is based on the remarkable symbiotic association between certain ants and some of the homopterous pest species. These ants, in exchange for the honeydew excreted by the homopterans (usually certain scale insects, mealybugs and aphids), tend them and protect them from attack by predators and parasites. By eliminating the ants from the infested tree or plant, either with a mechanical barrier or a tanglefoot trap around the base of the trunk or main stem, the pest population is left untended and thereby exposed to full attack by natural enemies. If the natural enemy is an effective one this results in a decline in the pest density.

There is no question of altered microenvironment or stimulated host plant here, but a question has been raised by the discovery that in certain ant-tended aphid or scale insect species, the ant directly stimulates the reproduction and population growth of the pests. To the extent that this also occurs, ant removal would result in a decline in pest abundance both through increased natural-enemy activity and reduced reproduction rate, making it difficult, if not impossible, to measure the true effect of the natural enemies.

The hand removal method. In certain cases where the pest population is restricted to the plant due to the sessile nature or weak dispersal powers of the species, the removal of all natural enemies from the experimental plant or plant part (e.g., branch) by hand has been attempted. This laborious technique was devised to circumvent the objections that stimulation of the pest population by the manipulation itself rather than the elimination of natural enemies was the cause of the difference between treatment and check. To carry out this method a group of workers participates on a continuous basis, night and day, in the hand removal of any and all natural enemies which settle onto the leaves and twigs of the protected tree or branch. This effort continues for the duration of the test, which must extend long enough to produce a real difference between treated and check plots.

The technique has been applied to avocado pests in southern California, including lepidopterous caterpillars, mites, mealybugs, and scales (Fleschner et al. 1955).

In many of these enemy-exclusion techniques the differences in density between the protected and check host populations have been substantial. In some demonstrated cases (e.g., citrus red scale, olive scale) the increased density of the protected host population has led to severe damage or destruction of the crop plant or plant part.

Exclusion studies in cotton. Versions of the exclusion methods just described have been used to study the effects of naturally occurring biological control agents in cotton. In one series of experiments, field-grown cotton was placed under large cages and then the plants in half the cages were treated with a short residue, broad spectrum insecticide to eliminate most of the insects, while those in the remaining cages were not treated and thus remained populated. Next, equal numbers of small bollworm *(Heliothis zea)* larvae or eggs were placed on the treated and untreated caged cotton plants, and after a lapse of time their fate was assessed. The analyses showed that where natural enemies (among the spectrum of insect species) were retained there was strikingly lower survival of bollworms than where the predators had been eliminated (van den Bosch et al. 1969).

In another type of experiment conducted in open fields, several insecticide programs for early- and mid-season control of *Lygus hesperus* were compared with untreated controls for their effects on natural enemies and subsequent infestations of late-season pests, namely the cabbage looper and the beet armyworm. Here the experiments were large enough (plot size was usually 40 acres and in one case 160 acres) so that interplot movement of natural enemies did not obscure the effects of the various insecticides. The studies clearly showed that where natural enemies were depressed by the *Lygus* control treatments there were substantial to striking increases in population levels of the lepidopterous pests over those in the untreated controls (Falcon et al. 1971).

The life table technique

An altogether different technique for evaluating the role of a natural enemy in influencing the population dynamics and mean density of a pest insect utilizes the life table method of data analysis. In this procedure, which at the present time is applicable mainly for pests with discrete, nonoverlapping generations, a tabular accounting of mortality and its cause or causes is determined for each insect stage over a sequence of from eight to fifteen or more generations of the pest population in a more or less sharply limited area. The data for each mortality factor so collected are analyzed statistically, or graphically, for correlation between pest density and changes in density and the amount of mortality resulting from each measured mortality factor. Such analysis will disclose the type of mortality for each factor, that is, whether density-dependent or density-independent, and if the former whether direct, delayed, or inverse, and the contribution of each type of mortality to population trends. A brief description of the methods of compiling and analyzing life tables for purposes of evaluating natural-enemy action in influencing pest population densities follows.

For life table construction a series of density samples of the successive life stages (eggs, larvae, pupae, and adults) is collected and the causes of mortality or decline in numbers during the life cycle are measured or inferred. An example of such a table, derived from LeRoux et al. (1963), is given in Table 2. In this table, the first column specifies the age or stage of develop-

Table 2. Life table for diamondback moth, *Plutella maculipennis* (Curtis) on cabbage, 1961, Ontario, Canada.*

Age or stage (x)	Density† (N_x)	Mortality factor (D_xF)	No. dying (D_x)	Percent of stage dying ($100\ D_x/N_x$)	Percent of generation dying ($100\ D_x/N_o$)
Egg	1580	Infertility	25	1.6	1.6
Larva I-IV	1555	1.20" Rainfall	1199	77.1	77.5
Larva	356	0.52" Rainfall	36	10.0	79.7
IV	320	*Microplitis plutellae*	52	14.6	83.2
Cocoon (prepupa)	268	*Horogenes insularis*	69	25.7	87.5
Cocoon (pupa)	199	*Diadromus plutellae*	92	46.2	93.3
Adult	107	Sex ratio corr.‡	1	0.9	93.4
Adult	106	Abnormal fecundity§	78	73.6	98.3
Adult (normal)	28				

Generation mortality: 99.5%

Expected eggs: 28/2 x 216 = 3024
Actual eggs: 864
Trend index: 864/1580 x 100 = 54.7%

*Modified from Harcourt (1963).
†Sampling unit: crown quadrant.
‡Sex ratio corr.: Actual SR was 54 males to 53 females. To reduce this to 50:50 requires correction of 1% (one male in 107 adults).
§Abnormal fecundity: Realized fecundity of a sample of reared females was only 26.4% of normal (216).

ment (x) of the pest, the next column the density or number per unit of habitat (N_x) of each of the successive age groups or stages, the third column specifies the various mortality factors (D_xF) acting on each age group, the fourth column lists the numbers killed (D_x) by the corresponding mortality factor, the fifth column gives the percentage dying ($100\ D_x/N_x$) relative to the numbers present at the beginning of the stage, and the last column lists these numbers killed as percentage mortality ($100\ D_x/N_o$) based upon the number of individuals (eggs) at the beginning of the generation. At the bottom of the table, departures in the actual sex ratios of emerging adults from the expected ratio of 1:1 male to female are entered as adjustments, either positive or negative, according to whether a lower or higher proportion of males was produced. Losses in the adult stage due to dispersal, mortality, reduced fecundity deriving from less than average size (the opposite of this, enhanced fecundity due to greater than average size would appear as a negative mortality correction) are listed subsequently. The total generation mortality, the expected number of eggs to be produced in the next generation, the actual number of eggs, and the index of population trend $[I = E(n + 1)/E(n) \times 100]$ complete the table.

The index of population trend provides an estimate of the expected density of the egg stage of the next host generation. In some life tables, the actual egg density of the following generation is used to compute this value, as in Table 2, in which case it specifies the actual population trend rather than the expected. Differences between the expected and actual trend index are considered to reflect immigration (or emigration) of adults into (or out of) the study site.

Thus far, a life table has been explained as a sequence of declining densities of the various stages in a generation, these declines being produced by mortality factors which impinge on the species being studied. The problem next to be considered is the manner in which mortality *changes* with pest density from generation to generation. To attack this problem, a series of life tables must be compiled, one for each generation in a sequence of consecutive generations. It is presumed, and experience suggests that the presumption is justified, that in a sequence of generations, ranging from eight to fifteen or so, the density of the subject species will rise and fall in a typical population cycle or fluctuation. Such a cycle will cover a range of densities from high to low, and in this range of densities the mortality factors acting upon the pest species will exhibit patterns of response to host density. Such patterns, as has been mentioned above, may be unrelated to density, in which case the mortality factor is classed as density-independent, they may be immediately and positively related to density, in which case the mortality factor is classed as direct density-dependent, or they may be related to density but only with a time lag, in which case the cause of mortality is classed as delayed density-dependent. It is also possible to envision a mortality factor which causes a level of mortality directly but negatively related to host density, and such a factor is called an inverse density-dependent factor.

Experience indicates that all classes of the mortality factors described above can occur at one time or another, and in some cases even simultaneously, during a generation of an insect. Theory prescribes the necessary presence of at least one direct or delayed density-dependent mortality factor acting during such a population cycle of a host insect. Such mortality actions are considered necessary in the *regulation* of population numbers because they increase their lethal action as host density increases and, vice versa, decline in lethal action as host density declines. A sequence of life tables derived from properly designed sampling schemes will enable these various patterns to be ascertained. Where introduced natural enemies function as important controlling agents in respect to host density, this life table technique will disclose this capability.

There are several data-processing techniques for analyzing life tables; the correlation analysis technique of Varley and Gradwell (1960, 1963, 1971) is somewhat simpler, and will be described here.

From the life table data the logarithms of the numbers (density) of each stage are determined and a series of k values computed. The k value for a stage is the difference between the logs of the densities at the beginning and end of the stage. Also, a value K, representing total mortality in the generation, is computed as the difference in the logs of numbers of individuals at the beginning of the generation and finally surviving to adulthood. Then, the following equation holds, for a species with six stages or age groupings

$$K = k_1 + k_2 + k_3 + k_4 + k_5 + k_6 .$$

According to the Varley and Gradwell analysis, if, from a series of life tables, the different values of k_x are plotted against log (N_x) the points should define a line of positive slope if the factor causing the k_x mortality is density-dependent, a line of zero slope if the factor is density-independent, and a negative slope if the factor is inverse density-dependent.

The level of mortality produced by natural enemies is often not related to the density of the host stage attacked but rather to the host density one generation earlier. This is because the number of the searching natural enemy adults in the present host generation is derived from and reflects the density of the previous host generation. This delayed density-dependent mortality can be demonstrated by the improvement in the correlation between k and log of host density when k is plotted against log $(N_x)_{n-1}$ rather than against log $(N_x)_n$ where n and $n-1$ are the host densities in the present and next previous generations respectively.

Varley and Gradwell (1960) refer to "key factors" in their analyses of life table data. A key factor is a mortality factor whose effects on the host are the most strongly correlated with change in final host density. It is most easily determined by ascertaining in which stage k value is most strongly correlated with total K.

Huffaker and Kennett (1969) have warned that life table analyses, by themselves, based as they are on correlation rather than demonstrated cause and effect, cannot serve to disclose true natural-enemy effectiveness. Evaluation of effectiveness must be based in part at least on experimental evidence. Future evaluations of the importance of natural enemies in bringing about suppression of host pests will very likely rely on a combination of "check methods," or experimental evidence, and analytical procedures.

REVIEW AND RESEARCH QUESTIONS

1. Describe the sequence of operations followed in carrying out a biological control program.

2. How can you tell that a pest insect is exotic in origin? How can you determine its country of origin?

3. How are natural enemy importations into the United States controlled?

4. What purpose does the mass culture insectary serve in a biological control agency?

5. What are the various experimental ways in which one can evaluate the effectiveness of an established natural enemy?

6. What is the trend index in a life table?

7. How can the life table be used to analyze the effectiveness of a natural enemy?

BIBLIOGRAPHY

Literature cited

DeBach, P. and B. R. Bartlett. 1964. Methods of colonization, recovery and evaluation. In *Biological control of insect pests and weeds,* ed. P. DeBach, Chap. 14, pp. 412-426, London: Chapman & Hall.

Falcon, L. A., R. van den Bosch, J. Gallagher, and A. Davidson. 1971. Investigation of the pest status of *Lygus hesperus* in cotton in central California. *J. Econ. Entomol.* 64: 56-61.

Fleschner, C. A., J. C. Hall, and D. W. Ricker. 1965. Natural balance of mite pests in an avocado grove. *Calif. Avocado Soc. Yearbook* 39: 155-162.

Harcourt, D. G. 1963. Major mortality factors in the population dynamics of the diamondback moth, *Plutella maculipennis* (Curt.) (Lepidoptera, Plutellidae). *Mem. Entomol. Soc. Canada* 32: 55-66.

Hille Ris Lambers, D., and R. van den Bosch. 1964. On the genus *Therioaphis* Walker, 1870, with descriptions of new species. Zoologische Verhandelingen No. 68. 47 pp.

Huffaker, C. B., and C. E. Kennett. 1969. Some aspects of assessing efficiency of natural enemies. *Canad. Entomol.* 101: 425-447.

LeRoux, E. J., R. O. Paradis, and M. Hudon. 1963. Major mortality factors in the population dynamics of the eye-spotted bud moth, the pistol casebearer, the fruit-tree leaf roller, and the European corn borer in Quebec. *Mem. Entomol. Soc. Canada* 32: 67-82.

van den Bosch, R. 1968. Comments on population dynamics of exotic insects. *Bull. Entomol. Soc. Amer.* 14: 112-115.

van den Bosch, R., B. D. Frazer, C. S. Davis, P. S. Messenger, and R. Hom. 1970. *Trioxys pallidus*—an effective new walnut aphid parasite from Iran. *Calif. Agric.* 24(11): 8-10.

van den Bosch, R., T. F. Leigh, D. Gonzalez, and R. E. Stinner. 1969. Cage studies on predators of the bollworm in cotton. *J. Econ. Entomol.* 62: 1486-1489.

van den Bosch, R., E. I. Schlinger, J. C. Hall, and B. Puttler. 1964. Studies on succession, distribution and phenology of imported parasites of *Therioaphis trifolii* (Monell) in southern California. *Ecology* 45: 602-621.

van den Bosch, R., E. I. Schlinger, E. J. Dietrick, K. S. Hagen, and J. K. Holloway. 1959. The colonization and establishment of imported parasites of the spotted alfalfa aphid in California. *J. Econ. Entomol.* 52: 136-141.

Varley, G. C., and G. R. Gradwell. 1960. Key factors in population studies. *J. Anim. Ecol.* 29: 399-401.

Varley, G. C., and G. R. Gradwell. 1963. The interpretation of insect population changes. *Proc. Ceylon Assoc. Adv. Sci.* 18: 142-156.

Varley, G. C., and G. R. Gradwell. 1971. The use of models and life tables in assessing the role of natural enemies. In *Biological control,* ed. C. B. Huffaker, Chap. 4, pp. 93-112. New York: Plenum Press.

Additional references

DeBach, P., ed. 1964. *Biological control of insect pests and weeds.* Sec. IV. The introduction, culture and establishment programme, pp. 283-426. London: Chapman & Hall.

6

Factors Limiting Success of Introduced Natural Enemies

Since the initial success against the cottony-cushion scale in California, nearly 100 pest insects and weeds in many parts of the world have been completely or substantially controlled by imported natural enemies. Despite this gratifying record, most attempts in classical biological control have either met with total failure, or they have been only partially successful (Turnbull and Chant 1961; Turnbull 1967).

In seeking guidelines to improve upon this record, the initial impulse is to look to the highly effective programs and out of them devise some kind of formula which might lead to an improved rate of success in the future. But this is difficult because each program is unique, and the factors which lead to success in one may have little or no bearing on another. Furthermore, it is characteristic of the successful programs that very often the basic reasons for the favorable result are really not known; we simply conclude that we have colonized a highly effective natural enemy or enemies (i.e., they are fully adapted to the target species, vigorous, have excellent searching powers, distribute their progeny efficiently, etc.), under favorable environmental conditions.

But there are certain guidelines and practices which if followed or ignored can either enhance or hinder the chances for success in given programs. These guidelines have been learned mostly from programs which failed or fell short of the mark, but in which the reasons for failure were identifiable. Analysis of a number of such programs has indicated that there are certain critical factors which contribute to failure, and it is these factors that we will discuss in this chapter.

CHARACTERISTICS OF THE COLONIZED ENVIRONMENT

The colonized environment is never a duplicate of the habitat from which a natural enemy is obtained, and the degree to which the two differ is of crucial importance to the success or even establishment of the species to be introduced. In other words, in a given program the physical and biological characteristics of the colonized environment may differ substantially from or closely approach those of the natural enemy's native habitat. This fact bears importantly on the success of the introduced species.

It is generally assumed that the chances for establishment and effective performance of a natural enemy will be greater where the environment into which it is being introduced resembles that of the native habitat than where the two environments differ widely. For example, one would anticipate that the chances for success of a parasite obtained from the Nile Delta would be greater in the subtropical Imperial Valley of California than in northern Idaho. But it should also be emphasized that it is not only the physical nature of the environment that is important but its biotic characteristics as well. Thus, even where the climate of a colonized environment closely simulates that of a natural enemy's native habitat, the enemy may fail because the colonized area lacks a vital alternative host or foodstuff, because the enemy is not adapted to the target pest, or because indigenous hyperparasites may adapt to it and preclude effectiveness.

Certainly the mere presence of the natural enemy's host is no assurance that the enemy will become established or prosper. For one thing, the two have different spectra of requisites, and even though they might share some, the natural enemy may fail because the environment lacks a single requisite which is vital to it alone. This situation is reflected in Table 3, which lists cases in which characteristics of the colonized environment either precluded natural-enemy establishment or hindered their fullest performances.

There are, of course, other examples of this sort and other kinds of environmental factors which have precluded the establishment or effective performance of introduced natural enemies, but those listed adequately illustrate the nature of the problem.

Such physical and biological environmental factors are not the only obstacles to success in classical biological control. In fact, other factors such as the kinds of natural enemies colonized, the numbers colonized, and the circumstances under which colonizations are made, have in the aggregate perhaps been equally detrimental to classical biological control programs.

Table 3. Some natural-enemy species whose establishment was prevented or performance impaired by adverse factors in the colonized environments.

Environmental Factor	*Country*	*Natural Enemy*	*Host Species*	*Reference*
Adverse Climate				
Unfavorable humidity	Mexico	*Eretmocerus serius* Silv.	*Aleurocanthus woglumi* Ashby	Clausen 1958
	U.S. (California)	*Aphytis maculicornis* (Masi)	*Parlatoria oleae* (Colvée)	Huffaker, Kennett, and Finney 1962.
	U.S. (California)	*Aphytis lignanensis* (Masi)	*Aonidiella aurantii* (Mask.)	DeBach et al. 1955
Temperatures too high	U.S. (California)	*Bathyplectes curculionis* (Thom.)	*Hypera postica* (Gyll.)	Michelbacher 1943
	U.S. (California)	*Praon exsoletum* Nees, *Aphelinus asychis* Walk.	*Therioaphis trifolii* (Monell)	van den Bosch et al. 1964
	U.S. (central California)	*Trioxys pallidus* (Halliday) (French strain)	*Chromaphis juglandicola* (Kalt.)	Messenger and van den Bosch 1971
Temperatures too low	U.S. (Louisiana)	Several South American parasites	*Diatraea saccharalis* (Fabr.)	Clausen 1956
	U.S. (California)	*Metaphycus helvolus* (Comp.)	*Saissetia oleae* (Bern.)	Clausen 1956
	U.S. (southern California)	*Cryptolaemus montrouzieri* (Muls.)	*Planococcus citri* (Risso)	Clausen 1956
Voltinism	U.S. (interior areas of Calif. where host is even brooded)	*Metaphycus helvolus* (Comp.)	*Saissetia oleae* (Bern.)	Clausen 1956
Lack of alternative host	U.S.	*Paradexoides epilachnae* Ald.	*Epilachna varivestis* Muls.	Landis and Howard 1940
	Australia, France, Italy, Argentina, Brazil, and others	*Macrocentrus ancylivorus* Roh.	*Grapholitha molesta* (Busck)	Clausen 1958
Lack of foodstuff	U.S. (eastern)	*Tiphia* spp.	*Popillia japonica* Newm.	Clausen 1956
Lack of competitiveness	U.S. (California)	*Aphytis fisheri* DeBach	*Aonidiella aurantii* (Mask.)	DeBach 1965
Lack of synchrony with host	U.S. (eastern)	*Agathis diversus* (Mues.)	*Grapholitha molesta* (Busck)	Clausen 1956
	U.S. (eastern)	*Hyperecteina aldrichi* Mesn.	*Popillia japonica* Newm.	Clausen 1956

POORLY ADAPTED NATURAL-ENEMY SPECIES AND STRAINS

One of the main reasons for failure of classical biological control programs has been the introduction of natural-enemy species and strains which were poorly or not at all adapted to the hosts against which they were introduced. An illustration of this situation is the great number of natural enemies introduced into California against the black scale, *Saissetia oleae* (van den Bosch 1968). By 1955, thirty-eight species of natural enemies had been introduced against this pest from scattered places around the globe. Of these, fifteen became established, but today only one of them, *Metaphycus helvolus,* is of major importance as an enemy of the scale. Seemingly, for years entomologists literally collected whatever parasites or predators they found associated with *S. oleae* and allied species, and sent them to California in the hope that one might be the long sought effective natural enemy. During these years, parasites were shipped to California from such widely scattered places as Brazil and Tasmania when, in fact, the genus *Saissetia* is indigenous to Southeast Africa. It is hardly surprising, then, that less than half of the imported black-scale natural enemies became established, and that only one of them, a South African species, came to be of any importance.

S. oleae is only one of a number of pest species which have been "shotgunned" with natural enemies in this manner. We have only to look at programs involving such pests as the gypsy moth, the brown-tail moth, the oriental fruit moth, the California red scale, and the European corn borer to find examples of similar broad spectrum introductions.

In a number of cases, natural enemies of pest insects in one genus have been introducted against species in different genera. This occurred where *Opius* spp. ex the oriental fruit fly, *Dacus dorsalis* Hendel (a tropical species) were colonized on the United States mainland against cherry fruit flies, *Rhagoletis* spp. (temperate species), and when *Opius* spp. ex Mediterranean fruit fly, *Ceratitis capitata* (an African species), were colonized in California against the walnut husk fly, *Rhagoletis completa* Cresson (a Nearctic species). It is hardly surprising that none of these *Opius* became established.

Occasionally an exotic pest's native habitat cannot be determined, and in such cases guesswork is necessarily involved in natural-enemy procurement and introduction. This has occurred with certain of the mealybug and scale insect pests of citrus and has resulted in the considerable colonization of poorly adapted natural-enemy species. But guesswork is becoming increasingly rare as our knowledge of the insect faunas and insect systematics and distribution increases, and in the future guesswork should not contribute significantly to the introduction of poorly adapted species.

But the problem of introducing proper races or strains of natural enemies continues to confront biological control workers. Here again, either physical or biological adaptation may be involved. An interesting case history involving differing adaptation to climate by strains of a parasite has been recently recorded from California. In this case, which is detailed in Chapter 7, a strain of the parasite, *Trioxys pallidus,* imported from France against the walnut aphid, *Chromaphis juglandicola,* failed to perform effectively in the hot interior valleys of California. But a second strain of the same species imported from Iran has performed outstandingly in the California interior.

A case involving differential biological adaptation in intraspecific strains of a natural enemy has recently been reported from Canada. This concerns *Mesoleius tenthredinis* Morley, a parasite of the larch sawfly, *Pristiphora erichsonii* Hartig. In this situation a population of the sawfly had reduced the efficacy of the parasite by haemocytic encapsulation of the latter's eggs. A search of Europe revealed the occurrence in Bavaria of an *M. tenthredinis* population which is adapted to the resistant Canadian *P. erichsonii.* The Bavarian *M. tenthredinis* has now become established in Canada, and is effecting considerable parasitization in the area of colonization. Canadian entomologists are encouraged that it will eventually spread and take a heavy toll of the sawfly over a wide area (Turner and Muldrew).

In addition to the two cases just cited there have been a number of instances where introduction of better adapted intraspecific strains of natural enemies has increased the efficacy of biological control programs. Included have been strains of *Aphelinus mali* against woolly apple aphid in China, *Prospaltella perniciosi* Tower against San Jose scale in Switzerland, *Aphytis maculicornis* (Masi) against olive scale in California, *Comperiella bifasciata* Howard against yellow scale in California, and *Trichopoda pennipes* Fabricius against southern green stink bug in Hawaii.

The several cases just cited clearly point up the importance of the utilization of the proper intraspecific strains of natural enemies in classical biological control programs. Almost certainly a number of past failures have resulted from the importation of the wrong strains, and there is a clear need to reassess these failures on the premise that with renewed effort effective, adapted natural-enemy strains might be found.

THE MECHANICS OF NATURAL-ENEMY INTRODUCTION

When we reflect further upon past biological control efforts it becomes apparent that certain aspects of the mechanics of natural-enemy introduction have doomed many programs to failure. In some cases the persons or agencies involved in the introductions simply did not have the competence to undertake the projects. Then, too, there have been situations where facilities, manpower, or financial support were inadequate to permit the mounting of meaningful efforts. As a result, at times sufficient numbers of natural enemies could not be produced to assure their establishment, and in some cases promising species were lost in culture. But even where adequate facilities and skilled staff have been available, some entomophagous insects have proven to be extremely difficult to propagate in the insectary. For example, for years it was impossible to rear certain species of *Coccophagus* until S. E. Flanders (1937) discovered that the males develop as hyperparasites.

Currently we have in our insectary at Albany, California, an alfalfa weevil parasite, *Bathyplectes anurus*, which has a very complex dual diapause (i.e., it first passes a period of diapause as a mature larva in the host cocoon, and then, while still in the cocoon, passes a second period of diapause as an imago). Thus far all efforts to prevent induction of diapause in this species have failed, as have efforts to effect early termination of diapause. Thus, in the insectary only one generation of the parasite can be produced per year. This of course means that the increase of *B. anurus* in the insectary is a very slow process and as a result its intensive field colonization is hindered. With some natural-enemy species, problems with diapause, mating, and similar biological phenomena have so far defied solution and this has precluded insectary propagation of the species.

Another factor which at times has affected the success of natural-enemy introductions, has been indifference on the part of those involved in colonization of the species. This has largely occurred where persons or agencies normally involved in other aspects of insect control have been charged with the responsibility of colonizing parasites or predators shipped to them by one or another of the biological control agencies. It is rather understandable when someone basically involved in or philosophically oriented to some other control tactic does not take a deep interest in the care, colonization, and manipulation of natural enemies shipped to him by some remote laboratory or outside agency. On the other hand, it is inexcusable that we persist in utilizing such a practice in our biological control programs.

CONCLUSION

The preceding discussion reveals that a number of factors affect the success or failure of introduced natural enemies. To some it might seem that blind luck as much as any other factor has been the key to the outstanding successes. This, of course, is not true. Koebele went to Australia to search for enemies of cottony-cushion scale because background investigation directed him there. Overwhelming evidence pointed to Australia as the native home of the citrophilous mealybug, and effective natural enemies were indeed found there. A thorough and widespread search produced the Iranian strain of *Aphytis maculicornis* adapted to olive scale in California. In these cases the right natural enemies were sought in the right places and they performed with maximum effect in the colonized areas. On the other hand, with the failures or partial successes we are faced with an intriguing question: How many of these cases can be turned into outstanding successes such as that involving the walnut aphid in California and hopefully, too, the larch sawfly in Canada? The odds are that the record can be improved the second time around.

REVIEW AND RESEARCH QUESTIONS

1. Give an example of a biological control project where an earlier failure led to a later success.

2. Why is knowledge of the environmental conditions in both the habitat to be colonized and the native habitat of the natural enemy important in biological control?

3. Of what value are races or strains of natural enemies that are used in biological control?

4. Discuss the meaning of host specificity in parasitic insects. What implications are there for biological control work from host specificity?

5. Give an example of a biological control project in which the pest insect became resistant to an imported natural enemy.

BIBLIOGRAPHY

Literature cited

Clausen, C. P. 1956. *Biological control of insect pests in the continental United States.* U.S. Dept. Agric. Tech. Bull. 1139, 151 pp.

Clausen, C. P. 1958. Biological control of insect pests. *Ann. Rev. Entomol.* 3: 291-310.

DeBach, P. 1965. Some biological and ecological phenomena associated with colonizing entomophagus insects. In *Genetics of colonizing species,* ed.

H. G. Baker and G. L. Stebbins, pp. 287-306. New York: Academic Press, 588 pp.

DeBach, P., T. W. Fisher, and J. Landi. 1955. Some effects of meteorological factors on all stages of *Aphytis lignanensis,* a parasite of the California red scale. *Ecology* 36: 743-753.

Flanders, S. E. 1937. Ovipositional instincts and developmental sex differences in the genus *Coccophagous. Univ. Calif. Publ. Entomol.* 6(95): 401-432.

Huffaker, C. B., and C. E. Kennett. 1962. Biological control of the olive scale, *Parlatoria oleae* (Colvee), in California by imported *Aphytis maculicornis* (Masi) (Hymenoptera: Aphelinidae). *Hilgardia* 32: 541-636.

Landis, B. J., and N. F. Howard. 1940. Paradexoides epilachnae, *a tachinid parasite of the Mexican bean beetle.* U.S. Dept. Agric. Tech. Bull. 721. 31 pp.

Messenger, P. S. and R. van den Bosch. 1971. *Biological control,* ed. C. B. Huffaker. New York and London: Plenum Press.

Michelbacher, A. E. 1943. *The present status of the alfalfa-weevil population in lowland middle California.* Calif. Agric. Exp. Sta. Bull. 677. 24 pp.

Turnbull, A. L. 1967. Population dynamics of exotic insects. *Bull. Ent. Soc. Amer.* 13: 333-337.

Turnbull, A. L., and D. A. Chant. 1961. The practice and theory of biological control of insects in Canada. *Canad. J. Zool.* 39: 697-753.

Turnock, W. S., and J. A. Muldrew (In press). *Pristiphora erichsonii* (Htg), larch sawfly. In *Biological control programs against insects and weeds in Canada.*

van den Bosch, R. 1968. Comments on population dynamics of exotic insects. *Bull. Entomol. Soc. Amer.* 14: 112-115.

van den Bosch, R., E. I. Schlinger, J. C. Hall, and B. Puttler. 1964. Studies on succession, distribution and phenology of imported parasites of *Therioaphis trifolii* (Monell) in Southern California. *Ecology* 45: 602-621.

7

Analysis of Classical Biological Control Programs

Nearly 100 pest insects and weeds have been completely or substantially controlled by imported natural enemies (Table 4). Some of these have been of minor or localized status while others have been species of continental distribution and great economic importance. In certain cases success was attained with but modest effort while in others it came only after elaborate preparation, dogged perseverance over many years, and great expense. Among the pest insects, by far the greatest number of successes have been scored against homopterous species, particularly diaspine and lecaniine scales. Some researchers have suggested that scale insects are particularly amenable to biological control because, being sessile during much of their life cycle, they cannot escape or avert natural enemies and the colonies once found are vulnerable to maximum exploitation. But others point out that scale insects are particularly common pests of horticultural crops, such as citrus, and that the considerable success against them may simply reflect the greater emphasis placed on biological control of such orchard pests. There is probably some validity to each contention but this is perhaps irrelevant, for striking successes have also been scored against species of Coleoptera, Lepidoptera, Diptera, and Hymenoptera in a variety of situations. This is evidence enough that the chances for successful biological control exist across a wide spectrum of the major pest groups.

In biological control of weeds, there seems to have been disproportionate success against cacti of the genus *Opuntia.* But here again, this simply seems to reflect intensive efforts against this particular pest group, for analysis of the remaining cases listed in Table 4 reveals that a wide variety of weedy plant species has been successfully attacked by imported natural enemies. This would seem to indicate that under the right circumstances

Table 4. **Insect pests and weeds substantially or completely controlled by imported natural enemies.**

Part I: INSECTS

Pest species						
Scientific name	*Common name*	*Crop attacked*	*Place where controlled*	*Type of natural enemy*	*Degree of control*	*Reference*
Homoptera						
Acyrthosiphon pisum (Harris)	pea aphid	alfalfa	California Hawaii	parasite	S*	DeBach 1964
Aleurocanthus spiniferus (Quaint.)	spiny blackfly	citrus	Japan Guam	parasite	C†	DeBach 1964
Aleurocanthus woglumi Ashby	citrus blackfly	citrus	Cuba Mexico	parasite	C C	DeBach 1964
Antonina graminis (Maskell)	Rhodes grass scale	grasses	Texas	parasite	S	Schuster et al. 1971
Aonidiella aurantii (Mask.)	red scale	citrus	Greece California	parasite parasites	S S	DeBach & Argyriou 1967 DeBach 1971
Aonidiella citrina (Coq.)	yellow scale	citrus	California	parasite	S	DeBach 1964
Aphis sacchari Zhnt.	sugarcane aphid	sugar-cane	Hawaii	parasite & predator	S	DeBach 1964
Aspidiotus destructor Sign.	coconut scale	coconut & other palms	Fiji Mauritius Portuguese W. Africa Bali	predator parasite	C	DeBach 1964

* = substantial.
† = complete.

Pest species		Crop attacked	Place where controlled	Type of natural enemy	Degree of control	Reference
Scientific name	Common name					
Asterolecanium variolosum (Ratz.)	golden oak scale	oak	New Zealand Tasmania	parasite	S	DeBach 1964
Cavariella aegopodii (Scop.)	carrot aphid	carrot	Australia Tasmania	parasite	C	van den Bosch 1971
Ceroplastes rubens Mask.	red wax scale	citrus persimmon tea	Japan	parasite	C	DeBach 1964
Chromaphis juglandicola (Kalt.)	walnut aphid	walnut	California	parasite	S-C	van den Bosch 1971
Chrysomphalus aonidum (L.)	Florida red scale	citrus	Israel Seychelles Greece	parasite predator parasite	C S S	DeBach 1964 DeBach 1964 DeBach & Argyriou 1967
Chrysomphalus dictyospermi (Morgan)	dictyosperum scale	citrus	Greece	parasite	S	DeBach & Argyriou 1967
Eriococcus coriaceus Mask.	blue gum scale	Eucalyptus	New Zealand	predator	C	DeBach 1964
Eriosoma lanigerum (Hausm.)	woolly apple aphid	apple	New Zealand	parasite	C	DeBach 1964
Eulecanium coryli (L.)	vine scale	grapvine plum, etc.	British Columbia	parasite	S	DeBach 1964

Table 4. (Continued)

Pest species		*Crop attacked*	*Place where controlled*	*Type of natural enemy*	*Degree of control*	*Reference*
Scientific name	*Common name*					
Icerya aegyptiaca Dougl.	Egyptian mealybug or fluted scale	bread-fruit avocado citrus	Caroline Islands	predator	S	DeBach 1964
Icerya montserratensis Riley & Howard	fluted scale	citrus	Ecuador	predator	S	DeBach 1964
Icerya purchasi Mask.	cottony-cushion scale	citrus	California	predator & parasite	C	DeBach 1964
Ischnaspis longirostris (Sign.)	black thread scale	fruit & timber trees coconut palm	Seychelles	predator	S	DeBach 1964
Lepidosaphes beckii (Newm.)	purple scale	citrus	Mexico Texas	parasite	S	DeBach 1964
Nipaecoccus nipae (Mask.)	avocado mealybug (coconut mealybug in Bermuda)	citrus	Greece	parasite	S	DeBach & Argyriou 1967
		multiple hosts	Hawaii	parasite	S	DeBach 1964
Operophtera brumata (L)	winter moth	forest, shade, & fruit trees	Nova Scotia	parasite	C	Embree 1971

Pest species						
Scientific name	*Common name*	*Crop attacked*	*Place where controlled*	*Type of natural enemy*	*Degree of control*	*Reference*
Orthezia insignis Dougl.	greenhouse *Orthezia*	orna-mentals	Kenya	predator	S	DeBach 1964
Parlatoria oleae (Colvée)	olive scale	olive deciduous fruits	California	parasite	S-C	DeBach 1964 Huffaker & Kennett 1966
Perkinsiella saccharicida Kirk.	sugarcane leafhopper	sugar-cane	Hawaii	predator	C	DeBach 1964
Phenacoccus hirsutus Green	hibiscus mealybug	multiple	Egypt	parasite	S	DeBach 1964
Phenacoccus iceryoides Green	—	coffee	Celebes	predator	S	DeBach 1964
Phenacoccus aceris Sign.	apple mealybug	apple	British Columbia	parasite	C	DeBach 1964
Pinnaspis buxi (Bouche)	—	coconut & other palms	Hawaii & Seychelles	predator	S	DeBach 1964
Planococcus kenyae (LeP.)	coffee mealybug	coffee	Kenya	parasite	C	DeBach 1964
Pseudaulacaspis pentagona Targ.	white peach scale	mulberry	Italy	parasite	S	DeBach 1964
		mulberry papaya oleander etc.	Puerto Rico Bermuda	predator	S	DeBach 1964
		mulberry	Georgian S.S.R. (USSR)	parasite	S	Kobakhidze 1965

Table 4. (Continued)

Pest species		Crop attacked	Place where controlled	Type of natural enemy	Degree of control	Reference
Scientific name	Common name					
Pseudococcus spp.	mealybug	citrus	Australia	predator	S	DeBach 1964
Pseudococcus citriculus Green	–	citrus				
Pseudococcus comstocki (Kuw.)	comstock mealybug	apple	eastern Russia	parasite parasite	C C	DeBach 1964 Kobakhidze 1965
Pseudococcus gahani Green	*Citrophilus* mealybug	citrus	California Chile	parasite parasite	C S	DeBach 1964 DeBach 1964
Quadraspidiotus perniciosus Comst.	San Jose scale	apple & other fruits	Switzerland	parasite	S-C locally	Mathys & Guignard 1965
Saissetia oleae (Bern.)	black scale	citrus olive	California	parasite	S	DeBach 1964
		olive	Peru	parasite	S	DeBach 1964
Saissetia nigra (Nietn.)	nigra scale	ornamentals	California	parasite	S	DeBach 1964
Siphanta acuta (Wlk.)	torpedo bug planthopper	several hosts	Hawaii	parasite	S	DeBach 1964
Tarophagus proserpina (Kirk.)	taro leaf-hopper	taro	Hawaii	predator	S	DeBach 1964
Therioaphis trifolii (Monell)	spotted alfalfa aphid	alfalfa	California	parasite	S	DeBach 1964

Pest species		Crop attacked	Place where controlled	Type of natural enemy	Degree of control	Reference
Scientific name	Common name					
Trialeurodes vaporariorum (Westw.)	greenhouse whitefly	tomatoes etc.	Australia	parasite	S	DeBach 1964
Trionymus saccharis (Ckll.)	pink sugar-cane mealybug	sugar-cane	Tasmania	parasite	S	DeBach 1964
			Hawaii	parasite	S	DeBach 1964
Lepidoptera						
Cnidocampa flavescens (Wlk.)	oriental moth	shade trees	Massachusetts	parasite	S	DeBach 1964
Coleophora laricella (Hbn.)	larch casebearer	larch	Canada	parasite	S	DeBach 1964
Diatraea saccharalis (F.)	sugarcane borer	sugar-cane	West Indies	parasite	S	DeBach 1964
Grapholitha molesta (Busck.)	oriental fruit moth	peach	Canada	parasite††	S	DeBach 1964
Harrisinia brillians B&McD.	western grape leaf skeletonizer	grape	California	parasite	S	DeBach 1964
Homona coffearia Nietn.	tea tortrix	tea	Ceylon	parasite	C	DeBach 1964
Laspeyresia nigricana (Steph.)	pea moth	vegetables	British Columbia	parasite	S	DeBach 1964

††A native parasite.

Table 4. (Continued)

Pest species		*Crop attacked*	*Place where controlled*	*Type of natural enemy*	*Degree of control*	*Reference*
Scientific name	*Common name*					
Levuana iridescens B.B.	coconut moth	coconut	Fiji	parasite	C	DeBach 1964
Nygmia phaoerrhoea (Donov.)	brown-tail moth	deciduous forest, & shade trees	northeast U.S.	parasite	S	DeBach 1964
			Canada	parasite	C	DeBach 1964
Pieris rapae (L.)	imported cabbageworm	cruciferous crops	New Zealand	parasite	S	DeBach 1964
Plutella maculipennis (Curt.)	diamondback moth	cruciferous crops	New Zealand	parasite	S	DeBach 1964
Stilpnoptia salicis (L.)	satin moth	forest trees	U.S. British Columbia	parasite	S	DeBach 1964
Coleoptera						
Anomala orientalis Waterh.	oriental beetle	sugarcane	Hawaii	parasite	S	DeBach 1964
Anomala sulcatula Burm.	—	sugarcane	Saipan	parasite	C	DeBach 1964

Pest species		Crop attacked	Place where controlled	Type of natural enemy	Degree of control	Reference
Scientific name	Common name					
Brontispa longissima selebensis Gestro	coconut leaf miner	coconut	Celebes	parasite	S	DeBach 1964
Brontispa mariana Spaeth	Mariana coconut beetle	coconut	Mariana Islands	parasite	S	DeBach 1964
Gonypterus scutellatus Gyll.	eucalyptus weevil	eucalyptus	S. Africa	parasite	C	DeBach 1964
			N. Zealand	parasite	S	DeBach 1964
			Mauritius	parasite	C	DeBach 1964
			Kenya	parasite	S	DeBach 1964
			Madagascar	parasite	S	DeBach 1964
Hypera postica (Gyll.)	alfalfa weevil	alfalfa	California: S.F. Bay Area	parasite	C	DeBach 1964
			San Joaquin Valley		S	DeBach 1964
			parts of eastern U.S.	parasite	S	van den Bosch 1971
Oryctes tarandus Ol	rhinoceros beetle	sugarcane	Mauritius	parasite	S	DeBach 1964
Promecotheca papuana Cziki	—	coconut	New Britain	parasite	S	DeBach 1964
Promecotheca reichei Baly	coconut leaf-mining beetle	coconut	Fiji	parasite	C	DeBach 1964
Rhabdoscelus obscurus (Bdv.)	sugarcane weevil	sugarcane	Hawaii	parasite	S	DeBach 1964

Table 4. (Continued)

Pest species						
Scientific name	*Common name*	*Crop attacked*	*Place where controlled*	*Type of natural enemy*	*Degree of control*	*Reference*
Diptera						
Ceratitis capitata (Wied.)	Mediterranean fruitfly	many fruits	Hawaii	parasite	S	Clausen et al. 1965
Dacus dorsalis Hendel	oriental fruitfly	many fruits	Hawaii	parasite	S	DeBach 1964
Dasyneura pyri (Bouche)	pear leaf midge	pear	New Zealand	parasite	S	DeBach 1964
Orthoptera						
Oxya chinensis (Thumb.)	Chinese grasshopper	sugarcane	Hawaii	parasite	S	DeBach 1964
Hymenoptera						
Cephus pygmaeus (L.)	European wheat stem sawfly	wheat	Ontario (Canada)	parasite	S	DeBach 1964
Gilpinia hercyniae Htg.	spruce sawfly	spruce	Canada	parasite	S-C	DeBach 1964
Pristiphora erichsonii (Htg.)	larch sawfly	larch	Canada	parasite	S	DeBach 1964

Pest species		Crop attacked	Place where controlled	Type of natural enemy	Degree of control	Reference
Scientific name	Common name					
Hemiptera						
Nezara viridula (L.)	green tomato bug or southern green stink bug	vegetables	Australia Hawaii	parasite parasites	S S-C	DeBach 1964 Davis 1967

Part II: WEEDS

Weed species		Place where controlled	Type of enemy	Degree of control	Reference
Scientific name	Common name				
Opuntia stricta Haw.	prickly pear cactus	Australia	pad-boring caterpillar	C	Nat. Acad. Sci. 1968
O. inermis D.C.	prickly pear cactus	Australia	pad-boring caterpillar	C	,,
O. imbricata (Haw.) D.C.	walking stick cholla	Australia	pad-feeding mealybug	C	,,
O. vulgaris Mill.	prickly pear cactus	Australia	pad-feeding mealybug	S	,,
O. streptacantha Lem.	prickly pear cactus	Australia	pad-feeding mealybug	S	,,

Table 4. (Continued)

Weed species		Place where controlled	Type of enemy	Degree of control	Reference
Scientific name	Common name				
O. megacantha Salm-Dyck	mission prickly pear	Australia	pad-feeding mealybug pad-boring caterpillar trunk borer	S	”
O. dilenii (Ker. Gawl) Haw.	prickly pear	Ceylon	pad-feeding mealybug	S	”
	prickly pear	New Caledonia	pad-boring caterpillar	S	”
O. tuna Mill.	prickly pear	Mauritius	pad-feeding mealybug pad-boring caterpillar	S	”
O. tricantha Sweet	prickly pear	West Indies	pad-boring caterpillar	S	”
Opuntia sp.	prickly pear	S. India	pad-feeding mealybug	S	”
Hypericum perforatum Linn.	Klamath weed or	California	leaf-feeding beetles & a root borer	C	”
	St. Johnswort	Australia	a leaf-feeding beetle	S	”
		Chile	a leaf-feeding beetle	S	”
Clidemia hirta L.	curse	Fiji	a leaf-feeding beetle	C	”

Weed species		Place where controlled	Type of enemy	Degree of control	Reference
Scientific name	Common name				
Cordia macrostachya (Jaquin) Roem. & Schult.	black sage	Mauritius	leaf- and seed-feeders	S	Nat. Acad. Sci. 1968
Lantana camara var. *aculeata* (L.) Moldenke	Lantana		a leaf-feeding bug & a leaf-feeding beetle	S	"
Eupatorium adenophorum (Spreng.)	Pamakani	Hawaii	gall-forming fly	C	"
Emex spinosa (L.) Campd.	spiny emex	Hawaii	leaf-feeder	S	"
E. australis Steinh.	spiny emex	Hawaii	stem-borer	S	"
Tribulus terrestris L.	puncturevine	Hawaii (Kauai)	seed-feeding beetle	C	"
T. cistoides L.	Jamaica fever plant	Hawaii (Kauai)	stem-boring beetle	C	"
Senecio jacobaeae L.	tansy ragwort	limited areas of N.W. California & central Oregon	leaf-feeding lepidopteran	S	Hawkes 1968
Alternanthera spp.	alligatorweed	parts of southern Georgia & northern Florida	leaf-feeding beetle	S	L.A. Andres pers. comm.

many additional exotic weed species can be considered as likely prospects for biological control.

Space limitation precludes analysis of each of the cases listed in Table 4, and in many cases, there simply are not enough data to permit meaningful analyses. Instead, we have elected to discuss in this chapter a limited number of cases which illustrate certain problems, approaches, or phenomena that have recurred in classical biological control programs.

THE WALNUT APHID

There are several noteworthy aspects to this highly successful program, namely: (1) it involved a pest that should have been controlled over half a century ago, (2) control was accomplished by a single natural enemy, the aphidiine wasp, *Trioxys pallidus,* (3) it involved deliberate use of an ecologically adapted strain of the wasp, and (4) it is a case in which initial failure was followed a decade later by virtually complete success. (See Figure 13.)

Figure 13. Female *Trioxys pallidus* attacking walnut aphids. Note that in the center of the photo there are two old aphid mummies from which wasps have emerged. Photograph by Ken Middleham, University of California, Riverside.

Chromaphis juglandicola has been a serious pest of walnut, *Juglans regia,* in California since the early part of the current century. For decades no thought at all was given to its biological control, but instead its annual outbreaks in thousands of acres of walnut were routinely treated with chemical insecticides. Initially, nicotine sulfate was the most widely used material but since the middle 1940s it has been supplanted by a variety of synthetic organochlorines and organophosphates. Chemical control of the aphid is costly, disturbing to the walnut ecosystem, causing rapid resurgence of the pest and the creation of "secondary pest" problems, and hazardous to warm-blooded animals. Furthermore, the aphid has developed resistance to a succession of insecticides used against it. One wonders, then, why biological control was not attempted many years ago. The answer simply seems to be that no one gave it serious thought. The aphid was in the groves virtually from the time that commercial walnut production was undertaken in California. Apparently most persons involved in walnut production and walnut pest control routinely attacked it with insecticides, giving little thought to its possible permanent suppression by biological control. It was no help either that for years (the woolly apple aphid case notwithstanding) there was a widely held belief that aphids were poor prospects for biological control presumably because their high reproductive capacities and ability to develop at low temperatures gave them an insurmountable advantage over natural enemies.

The idea that aphids were not generally susceptible to classical biological control was completely dispelled by the striking success scored against the spotted alfalfa aphid in California in the middle 1950s (van den Bosch et al. 1964.) This success turned thoughts to other aphid species as possible targets of biological control, a particularly ripe field since virtually all of California's pest aphids are exotic species. Among these, the walnut aphid was selected as a prime target. There were several reasons for this: (1) in California it is monophagous and thus restricted to the walnut habitat, which eliminated concern over the need for finding parasites adapted to both primary and alternative host plants and their habitats, (2) the walnut ecosystem is a stable, relatively long-lived system and thus favorable to the establishment of a meaningful host-parasite relationship, (3) *C. juglandicola* was obviously an insect suitable for attack by an effective natural enemy in that it was generally and perennially epidemic and its populations were free of significant attack by competing parasites, i.e., the parasite niche was vacant, (4) related Callaphididae (including the spotted alfalfa aphid) were known to have important parasites in the Old World.

Walnut (*J. regia*) is a Palaearctic species whose original native habitat extends from southeastern Europe to China (Bailey and Bailey 1949). But over centuries man has widely distributed this prized tree, so that today it is found on all continents, and in the Old World it occurs from the Atlantic Seaboard to Europe to the Pacific Orient.

The universal distribution of walnut and the walnut aphid in the Old World contributed to the initial introduction of a poorly adapted strain of *Trioxys pallidus* into California. The basic mistake lay in the collection of the parasite near Cannes in southeastern France. There was no particular reason for selecting this area, excepting that its climate is in some respects similar to that of parts of California, and a University of California collector happened to be in the area. It was assumed that perhaps once introduced into California genetic selection in the field would enable the parasite to disperse into and become effective in all of the walnut-growing areas. At any rate, the parasite was readily collected at Cannes and shipped to California in the spring of 1959. After quarantine processing it was passed on to the insectary for mass propagation (van den Bosch et al. 1962). In the insectary large-scale propagation of the parasite was quickly attained and sizable field colonizations were possible even before the end of 1959. In subsequent years tens of thousands of the French wasps were colonized in virtually all walnut-growing areas of California.

The French *Trioxys pallidus* was very quickly established at several localities in coastal southern California, and rapidly increased in abundance at these places. Studies in San Diego County showed that the parasite was capable of destroying a very high percentage of the aphid population in areas of mild-equable climate. But this is a marginal walnut-growing area; major production occurring in the hot-arid interior, particularly the Great Central Valley. In this valley the French strain of *T. pallidus* never established a foothold, despite the colonization of many thousands of wasps on heavy aphid populations at a variety of locations. For example, in the San Joaquin Valley (the southern portion of the Great Central Valley), where approximately 75,000 French *T. pallidus* were colonized, recoveries were not made beyond the seasons of colonization.

The disappointing performance of the French *T. pallidus* in the California interior was a clear indication that it lacked genetic characteristics which would enable it to thrive or even survive in areas of extreme summer heat and low humidity. Consequently, a decision was made to seek a new strain from an area climatically similar to the Great Central Valley. This search was conducted on the central plateau of Iran, which has a summer climate nearly identical to that of the San Joaquin Valley.

Trioxys pallidus was obtained from Iran in the late spring of 1968, and colonized in several areas of central California during the summer and autumn of that year. Results were prompt and spectacular. Even though only small colonizations were made at the most unfavorable time of year (midsummer), recoveries were made at all colonization sites. Furthermore, an autumn survey at one of the colonization sites showed that the parasite had effected a high level of parasitization and had spread at least a mile from the colonization focus. The wasps then survived the winter of 1968-69 and continued to increase in abundance and to disperse. Additional colonizations were made in 1969, and establishment was again invariably effected. By the autumn of 1969 *T. pallidus* was widely established in the Great Central Valley and in several valleys peripheral to San Francisco Bay. Surveys conducted in the spring and early summer of 1970 showed that the parasite was having a heavy impact on *C. juglandicola* at a number of places. By the end of that year it had virtually covered all of the walnut-growing areas of central and northern California, an area of perhaps 50,000 square miles (van den Bosch et al. 1970). In the spring of 1971, *T. pallidus* had a generally crushing impact on the aphid. Surveys indicated that economically injurious infestations of the aphid were virtually nonexistent in the major production areas. In two monitored groves parasitization of the fundatrix generation (aphids hatching from overwintering eggs) exceeded 90 percent. This degree of parasitization appeared to be general in commercial groves.

All of this recent activity of *T. pallidus* is attributable to the Iranian strain which is quite obviously fully adapted to the hot interior of California. After four years of vigorous activity it must be assumed that the Iranian *T. pallidus* will continue to be a major population determinant of *C. juglandicola.* Already the wasp can be credited with substantial commercial control of this aphid. The ecological benefit resulting from diminished insecticide usage is an additional bonus being derived from the parasite. But perhaps the greatest benefit resulting from the walnut aphid program is the indisputable demonstration of the importance of using adapted intraspecific strains of a natural enemy in biological control programs. This is a lesson that should not be overlooked in future programs.

THE WINTER MOTH

This program is especially notable because the two parasites which effected essentially complete control of the winter moth in Nova Scotia and neighboring eastern provinces in Canada were not particularly prominent species in their area of indigeneity (Europe). The important implication here is that it is a serious mistake to prejudge imported natural enemies on the basis of their performance in their native habitats.

Operophtera brumata is a European moth (Geometridae) whose larvae feed on the foliage of hardwood species and thus pose a serious threat to forest, ornamental, and orchard trees. The pest was accidentally introduced into Nova Scotia some time in the 1930s, and by the middle 1950s it had covered approximately one-third of that province, occurring in moderate to severe infestations over several hundred square miles. In a span of ten years in just two counties the moth destroyed 26,000 cords of oak wood valued at approximately $2 million. Furthermore, if its depredations had remained unchecked it quite possibly would have wiped out all of the oak stands in Nova Scotia, caused severe damage to other forest species, shade, and orchard trees, and spread epidemically over much of Canada. The introduced parasites have unquestionably slowed the spread of the winter moth, and even if it moves into new areas they appear capable of maintaining the pest at low, endemic levels.

Parasite introductions against *O. brumata* were initiated in 1954, and of six species colonized, two, the tachinid *Cyzenis albicans* (Fallen) and the ichneumonid *Agrypon flaveolatum* (Gravenhorst) became established. By 1965 essentially complete biological control of *O. brumata* had been effected, with *C. albicans* being largely responsible for the pest's population collapse.

The winter moth program is one of the most meticulously analyzed biological control programs on record. The study is a particularly outstanding example of the utility of the life table technique in demonstrating the crucial role of parasites in host population regulation. For example, in 1958 in a plot at Oak Hill, Nova Scotia, before parasitization had reached 10 percent, a total generation mortality rate of 96.5 percent was recorded, indicating an increasing population. But in 1960, when parasitization had reached 72 percent the overall mortality percentage was 99.7, indicating a decreasing population. In each case the trend indicated by the life table was confirmed by data taken the following year. Thus, in 1959 the population density showed a twofold increase, while in 1961 it decreased by tenfold.

Analysis of thirty-seven life tables which were compiled during the study permitted the development of a population model which showed that parasitism had indeed become the key factor in control of the winter moth. Thus, before the parasites rose to significant status, generation survival was correlated with egg and early instar survival. However, after parasitization increased beyond 10 percent this relationship was completely reversed so that generation survival was correlated with late larval and pupal survival, rather than with early instar survival. In addition pupal survival was correlated with pupal parasitization. Embree (1971), the principal investigator, quite forthrightly points out that "the success in this experiment can be described as mere good fortune," because the choice of the six parasites out of sixty-three species known to attack *O. brumata* for introduction into Nova Scotia was largely a matter of chance. Furthermore, the outstanding performance of *Cyzenis* and to a lesser extent *Agrypon* could not have been anticipated on the basis of their performance in Europe. On the other hand, the very careful studies of Embree and his group clearly reveal why the two parasites performed so effectively once they had become established in Canada. The key to this success is that the two species are compatible and complementary; one, *Cyzenis,* is effective at high host densities, the other, *Agrypon,* at low densities. Thus, as host abundance declines so does the efficacy of *Cyzenis,* but then at the lower densities *Agrypon* becomes relatively more efficient and in this way complements the former. This case, and the case of olive scale in California, cited below, illustrate the conspicuous, beneficial, joint action which very commonly results from the introduction of more than a single "best" species.

THE ORIENTAL FRUIT FLY

This program is particularly noteworthy because of the very striking successional sequence of the three key parasites involved. It provided an excellent opportunity for study of the competitive mechanisms of parasitoids within the host. The case is additionally significant because, as in the one involving the eucalyptus snout beetle (described below), a parasite which attacks the earliest host developmental stage (egg) ultimately prevailed and effected a significant control.

Dacus dorsalis, as its common name implies, is an Asiatic species which invaded Hawaii, probably via military transport, sometime in the 1940s. Under the salubrious Hawaiian climate and with abundant host fruits available to it the fly quickly erupted to enormous abundance and by the middle 1940s became a severe pest of a variety of fruits grown in the islands.

The severe local problem caused by *D. dorsalis* in Hawaii and the threat it posed to mainland agriculture engendered an intensive control and quarantine program by a consortium of institutions and agencies consisting of the Hawaii Board of Forestry and Agriculture, the University of Hawaii, the U.S. Department of Agriculture, the California State Department of Agriculture, and the University of California Agricultural Experiment Station. Biological control was one of the earliest and most intensive control efforts undertaken by this group. Explorers were sent to virtually every tropical and subtropical area on earth, while receiving, propagation, colonization, and evaluation teams were assembled in Hawaii. Actual parasite introduction was initiated in 1947-48 by the Hawaii Board of Forestry and Agriculture, whose collectors shipped parasitized fruit fly material from the Philippines and Malaysia.

Among the parasites obtained over the course of the program, three species, *Opius longicaudatus* (Ashmead), *O. vandenboschi* Fullaway, and *O. oophilus* Fullaway, played significant roles in the biological control of the pest, with the latter eventually dominating (Clausen et al. 1965). By sheer chance the sequence of establishment of the three parasites was such that as time passed a pattern of succession unfolded, in which the potentially least effective one (*O. longicaudatus*) first flourished and then was replaced by the next most promising species (*O. vandenboschi*), which in turn was displaced by the most promising one (*O. oophilus*).

The basis for this pattern lay in the ease of propagation of the parasites, and thus in the number available for field colonization. *Opius longicaudatus* was the most easily propagated and hence the most heavily colonized, and the first to become widely established. This all came about because of the prevailing ignorance of the biologies of the three species at the time they were introduced. At that time it was assumed that all three attacked medium-sized or large fruit fly larvae, and so larvae of these sizes were offered to the wasps for oviposition. But in actuality only *O. longicaudatus* attacks larger fruit fly larvae, while *O. vandenboschi* attacks the first larval instar, and *O. oophilus* attacks the fruit fly egg.

It is really a wonder, then, that any *O. vandenboschi* and *O. oophilus* at all were propagated and colonized. In fact, establishment of the latter was simply sheer luck, because in being superficially similar to *O. vandenboschi* it was never recognized during the propagation and colonization phases of the program as a distinct species. It seems that quite by chance some of the infested fruits used in the parasite propagation program contained *D. dorsalis* eggs as well as partly developed larvae, and it was on these presumably that *O. oophilus,* mixed in with the *O. vandenboschi* stocks, reproduced itself.

The propagation and colonization history of *O. oophilus* can perhaps be considered an embarrassment to the entomologists involved in the Oriental fruit fly program, but it also teaches us a lesson about the great care that must be exercised in the introduction of natural enemies. In all fairness to those involved it should be pointed out that there were great pressures to colonize the parasites during the crisis stage of the oriental fruit fly program. Over the course of four years, fourteen explorers were in the field and they shipped 4½ million fruit fly puparia of more than sixty species to the Honolulu quarantine laboratory. These yielded over twenty species of parasites plus several predators. Ultimately a total of twenty-nine parasite and predator species were released in the islands. During the critical years 1947 to 1953, 1.1 million parasite and predator adults were produced in the insectary.

This was an overwhelming task for both the quarantine and insectary teams, and the accidental propagation and release of *O. oophilus* is thus more understandable. Fortunately, no undesirable species sifted through the screen, but we should be sufficiently chastened by the *O. oophilus* incident to make sure that no such happening recurs.

The imported parasites of *D. dorsalis,* especially *O. oophilus,* have had a striking impact on their host. All three parasite species were first liberated in the summer of 1948 and *O. longicaudatus* was recovered on the island of Oahu in October of that year, while *O. vandenboschi* and *O. oophilus* were first recovered about two months later. *Opius longicaudatus* increased rapidly, parasitizing up to 30 percent of the fruit fly larvae in wild guava by January 1949. This level of parasitization was sustained for several months thereafter. But then *O. vandenboschi* increased in abundance, and by October 1949 it had attained higher levels of parasitization than its predecessor. With the increase in importance of *O. vandenboschi, O. longicaudatus* rapidly faded to very low abundance. *O. oophilus* began to assert itself in the summer of 1950, and by the end of that year it was the overwhelmingly dominant species, parasitizing an average of 70-75 percent of the developing fruit fly larvae in wild guava by the end of the year. In 1951, *O. oophilus* gained full dominance in Oahu, and since then it has remained the dominant parasite of *D. dorsalis.* In fact its dominance is so great that over most of the Hawaiian Islands *O. longicaudatus* and *O. vandenboschi* have been relegated to the status of rare species, being, in effect, displaced.

O. oophilus has effected substantial biological control of *D. dorsalis.* Today the population level of the fly is strikingly lower than what it was before the parasites were introduced, and many crops, including avocado, banana, papaya, loquat, peach, and persimmon are no longer seriously affected by it. Even in mango, a favorite host, infestation levels seldom exceed 10 percent and this permits economical supplementary control with compatible chemical sprays.

The dominance of *O. vandenboschi* and *O. oophilus* over *O. longicaudatus* derives from their earlier hatching in hosts and the resultant inhibition of the eggs and larvae of *O. longicaudatus* which may occur in the same hosts. Furthermore, in larvae simultaneously parasitized by *O. oophilus* and *O. vandenboschi,* the former invariably prevents development of the latter. Thus, with the very high levels of parasitization attained by *O. oophilus* there is virtually no chance for *O. longicaudatus* and *O. vandenboschi* to successfully parasitize a significant proportion of the fruit fly population. This is the secret to the success of *O. oophilus.*

THE IMPORTED CABBAGEWORM

This biological control case is presented to illustrate two points: (1) that a lack of knowledge of the adaptability of a natural enemy to a particular host pest can lead, after much effort, to little benefit, and (2) that it is valuable to reassess partially successful biological control projects in the light of our present awareness of the critical importance of host specificity and host suitability in the effectiveness of imported parasite species (or intraspecific strains).

The imported cabbageworm is one of the principal pests of cole crops (cabbage, brussels sprouts, collards, kale, broccoli). This pest, long established in North America, originated in Europe where it coexists with another cabbage-attacking butterfly, the European cabbageworm, *P. brassicae* (Linnaeus).

In Europe, larvae of both species are attacked by the gregarious, endoparasitic braconid, *Apanteles glomeratus.* It will be recalled from our chapter on the history of biological control that *A. glomeratus* was the first parasitic insect to be recorded in the literature, when in the early seventeenth century it was described emerging from larvae of *P. rapae.* This same braconid also bears the distinction of being the first parasite to be introduced (by C. V. Riley) into North America (1883) from another country (England) for the biological control of an insect pest (the imported cabbageworm).

Numerous publications and reviews of biological control in the past twenty to thirty years refer to this biological control effort, stressing the historical role of the parasite, and generally referring to the effort as partially to moderately effective. However, up to and including the early 1960s, pesticide control of the imported cabbageworm was usually considered necessary in order to produce a commercially acceptable crop. In spite of its historical interest, *A. glomeratus* cannot really be credited with providing effective control of *P. rapae.*

However, our story is not ended. In fact, an important bit of detective work now unfolds. In 1956 a German expert in biological control, Hans Blunck, surveyed certain parts of the United States and Canada for natural enemies of the imported cabbageworm. He observed *A. glomeratus,* among other natural enemies, to be fairly generally distributed in the areas he searched, attaining levels of parasitism of up to 50 percent in certain California cole crop fields. However, he was surprised over his failure to discover another *Apanteles* species very common on *P. rapae* in Europe, namely *A. rubecula* Marshall. Blunck noted (1957) that in Germany *A. glomeratus* seldom parasitizes *P. rapae.* Rather, this host is very frequently attacked instead by *A. rubecula.* He indicated that *A. glomeratus,* on the other hand, is the most common parasite of the related host, *P. brassicae.* Richards (1940) earlier reported similar results in England.

Then, in the period 1963 to 1965, Canada Department of Agriculture entomologist A. T. S. Wilkinson, in making a survey of parasites of the imported cabbageworm in the south coastal areas of British Columbia, discovered the presence for the first time in North America of *Apanteles rubecula* (Wilkinson 1966). It was found well established, the most abundant of all parasites (others were two tachinids, but not *A. glomeratus*), and the most widespread. Wilkinson was unable to explain the origin of this parasite, but we can presume that it came to British Columbia accidentally, perhaps as cocooned pupae or as larvae in host caterpillars carried on cabbage conveyed in commercial aircraft or in maritime ships' stores.

Stressing again that *A. rubecula* is almost specific to *P. rapae,* Wilkinson recorded it to parasitize the host up to levels of 50 percent. It showed itself to be a solitary endoparasite, emerging from fourth-stage host larvae, rather than fifth-stage larvae as in the case of the gregarious *A. glomeratus.*

In 1967, after initiating a new project on the biological control of cole crop pests in Missouri, U.S. Department of Agriculture entomologists theorized that *A. rubecula* might very well be an important importation candidate for control of *P. rapae* (Puttler et al. 1970). At that time *P. rapae* in Missouri required pesticide applications to keep it in check, even though *A. glomeratus* was present there. Puttler and his associates (1970) reported some additional, important clues regarding the suitability of *P. rapae* as a host for *A. glomeratus.* First they cited Boese (1936), who reported that eggs of *A. glomeratus* are sometimes encapsulated after being oviposited into *P. rapae* larvae. Puttler and colleagues also found this to be the case in Missouri whenever host larvae larger than first instar were attacked. These findings, coupled with the evidence that *A. rubecula* eggs are not encapsulated in this host and that it is almost specific to *P. rapae* (Richards 1940), caused them to conclude that *A. glomeratus* is not closely adapted to this host species, but that *A. rubecula* is.

Several colonies of *A. rubecula* were imported from British Columbia to Missouri and colonized on *P. rapae* in 1967. At the same time the egg parasite *Trichogramma evanescens* Westwood, known to be naturally adapted to *P. rapae,* was imported from eastern Europe.

U.S. Department of Agriculture entomologist F. D. Parker noted that at the beginning of each growing season cabbageworm populations occurred in very low numbers, and were essentially unparasitized but then rapidly exploded to damaging levels, after which time parasitization rose to substantial levels. He decided to release large numbers of the newly imported natural enemies at the beginning of the season so as to promote a more rapid and effective buildup of the control agents (Parker 1971, Parker et al. 1971).

This attempt was only partially successful; parasitization did rise more rapidly but still not sufficiently to prevent damage. The host was so rare at the start of the season that the parasites had great difficulty in finding them before they "escaped." Parker reasoned that if there were more hosts during the spring, the parasites would be able to increase more effectively.

Release in the late spring of both *P. rapae* eggs and parasite adults (both species) provided just the right combination of numbers. The host started to build up as usual, but now the natural enemies increased even more rapidly. In spite of the addition of the very pest being controlled, the cabbageworm population was curtailed before damage occurred in the late summer and fall.

Thus, though the imported cabbageworm had been the subject of a biological control effort since 1883, efforts to obtain the properly adapted natural enemies associated with it in its native home were only attempted in the past four to five years. Information on the existence of the "true" parasite of *P. rapae,* together with evidence of the maladaptedness of the original import, *A. glomeratus,* had been available for more than twenty-five years (Boese 1936, Richards 1940). The lesson this holds is clear; each case of biological control must be carried out with great attention to all the subtle details of host and natural-enemy distribution, parasite adaptedness, and seasonal host-parasite synchrony, before success can be anticipated with confidence. Such efforts are never likely to be made unless an enthusiastic, well-trained, and clearly identified group of specialists are charged with the full direction of such programs and are given adequate funds to explore all possibilities. This has not commonly been the case for much of the biological control work done throughout the world.

THE EUCALYPTUS SNOUT BEETLE IN SOUTH AFRICA

This project constitutes a classic example of the control of an invader pest by means of a natural enemy imported from the pest's home grounds. It provides further an example of control by an egg parasite, a type of natural enemy not previously considered to be particularly effective.

The scene of this accomplishment is South Africa, the protected plant, the eucalyptus tree, a native of Australia. As in the southwestern United States, so in South Africa, eucalyptus during the nineteenth century was imported from Australia to serve as a rapid-growing, potential hardwood timber source. As in the United States, the establishment of eucalyptus was accomplished to the exclusion of any of the multitude of serious pests which attack it in its original home. So for many years eucalyptus was spread throughout those parts of South Africa suitable to it and plantation stands and windbreaks thrived.

Then, in 1916, a pest of eucalyptus, identified as *Gonipterus scutellatus* Gyllenhal, was discovered near Cape Town. In the following years the weevil, whose adults and larvae feed on leaves and new shoots, spread rapidly, and it soon became apparent that "the snout beetle was threatening the entire eucalyptus-growing industry of the country" (Tooke 1953).

The expert identification of the pest provided the evidence of its origin. All of the members of the curculionid genus *Gonipterus* originate in Australia. *G. scutellatus* may be said to have coevolved with *Eucalyptus.*

In 1925, South African entomologists initiated a biological control program against the pest. For several years prior to this time it was realized that in the native home of the snout beetle, Tasmania and Australia, it was not considered a pest, and was not prominent. In fact its biology and ecology were little known. The entomologists concluded from this state of affairs that the snout beetle was not a damaging pest in its native home, and that this might be due to the presence in Australia of effective natural enemies. So, in 1926, entomologist F. G. C. Tooke was sent to Adelaide, South Australia, to search for natural enemies, to observe the economic status of the snout beetle, the species of *Eucalyptus* attacked, and the climatic conditions prevailing in the infested areas.

Since Australian entomologists were uncertain as to where snout beetle infestations might be found, Tooke proceeded to host tree stands to search foliage for signs of weevil damage. Within a month he found egg masses of the pest on eucalyptus leaves in southeastern coastal South Australia. Dissection showed that some of the eggs contained in the egg capsules were blackened, a sign of egg parasitism which proved to be due to a mymarid wasp, later named *Anaphoidea nitens* Girault (*Patasson nitens*). He collected and shipped some of these egg capsules to South Africa and later several tachinid parasites were also discovered and they too were shipped to South Africa.

Tooke considered the egg parasite to be quite promising, which it later proved to be. Collections showed it to be the most abundant natural enemy of the pest. Early season egg parasitization rates reached nearly 70 percent. Many rearings disclosed no hyperparasites. He did encounter a shipping problem, as the developmental period of the egg parasite was found to be shorter than the three-week duration of the sea voyage from Adelaide to Durban, South Africa. This required cold-storage shipment, necessitating a study of the consequences of cold exposures to the parasite.

The first shipments of *A. nitens* were received in good condition in South Africa in 1926. Colonization was effected promptly and successful establishment was observed in 1927. During the next four years a large-scale insectary propagation program aided in the dissemination of the parasite. Rapid natural dispersal also occurred.

The results of this importation were dramatic. Complete control occurred within three years in climatically suitable areas of South Africa. Substantial control was attained on most eucalyptus species in all the rest of the Union where the trees were grown. By 1941 economic control was produced on all but one eucalyptus species in the climatically severe regions.

The effectiveness of the egg parasite was unexpected. Conventional belief among biological control experts was that egg parasites were not promising control agents. This attitude was based partly on the rather poor showing attained up to that time from the extensive work with *Tricho-gramma* egg parasites, and the fact that no previously successful biological control case was due to an egg parasite. It was also felt that most pests susceptible to biological control were heavily attacked in the larval stages by parasites so that any egg parasitism produced mortality would have occurred in any case later in the life cycle. Egg parasites were thought less dependable, too, by virtue of their being considered more nonspecific and polyphagous in habit.

The success of the eucalyptus snout beetle program proved that an egg parasite can be an effective control agent. It also confirmed what a number of other successful biological control cases have since demonstrated, that effective control could be accomplished by just one parasite species. This, too, was not anticipated since the logical control theory at the time held that a sequence of natural enemies attacking the successive stages of a pest during its life cycle was more promising of control than merely a single enemy.

THE OLIVE SCALE IN CALIFORNIA

This project illustrates two important elements of biological control. One of these is the existence of strains or races of certain natural-enemy species, some of which turn out to be ineffective, while others prove to be effective; the problem is to recognize and perhaps capitalize on this situation. The other unique element in this project is the complementary action of two parasites, both attacking the same host population.

Olive scale was the most serious pest of olive in California from soon after the time of its discovery in 1934 until after a biological control campaign was begun in 1948 (Huffaker et al. 1962). Besides olive, the polyphagous scale attacks some 200 other host plants, including peach, apricot, plum, almond, and numerous ornamentals. On olive, the pest not only attacks twigs and leaves, but also fruits.

During the period before biological control was accomplished, commercially acceptable crops of olives could not be produced without use of pesticide applications.

Parlatoria oleae Colvée is believed native to northern India and Pakistan, though it has been known from the Mediterranean region for centuries. Initial searches for natural enemies were made in the Mediterranean countries, since the solitary ectoparasite *Aphytis maculicornis* had been known to be associated with olive scale in Italy since 1911. First importations were of *A. maculicornis* from Egypt. This strain became established by 1949, though parasitization rates by it were never high. In subsequent years additional imports of *A. maculicornis* were made from India, Iran, and Spain.

Careful study of the life histories of these different imports established that because of certain biological differences each could be classed as a "strain" or race (Hafez and Doutt 1954). The original Egyptian stock exhibited a thelytokous mode of reproduction (no males known, females produce all female offspring by parthenogenesis). The Spanish strain was deuterotokous, virgin females producing mostly female offspring (but with some males) parthenogenetically. The Iranian and Indian strains were both arrhenotokous, offspring of both sexes being produced only after mating, males only being produced parthenogenetically by unmated females. Differences also existed in the rates of development of these strains.

Following colonizations of the Egyptian strain in 1948 to 1949, the Persian, Indian, and Spanish strains were released in 1952 to 1953. Only the Persian colonizations increased in numbers to high levels. The assumption made later was that in all probability only this strain possessed the required adaptation to persist at moderate levels in California environments.

During the period 1952 to 1960, millions of the Persian *Aphytis* were colonized throughout much of California. Establishment was soon verified in many localities. But the degree of biological control was variable and in many cases not commercially adequate. The main reason for this is the harm caused to *A. maculicornis* by the intensive summer heat and low humidity which occurs in the commercial olive-growing areas of California.

Nevertheless, in some olive groves, very satisfactory commercial control of the scale was provided by *A. maculicornis.* Furthermore, subsequent to the establishment of this parasite there followed a marked general decline in the pest on many of its other host plants.

During the importation efforts of 1951 to 1952 which led to the discovery and successful establishment of the Iranian strain of *A. maculicornis,* several other species of parasitic Hymenoptera were reared from imported olive scale samples in the quarantine laboratory. Several of these proved to be hyperparasitic, and hence were exterminated. One turned out to be an endoparasitic species of *Coccophagoides,* which because of complications in its life cycle proved difficult to rear in the laboratory. The culture of this parasite was subsequently lost (Broodryk and Doutt 1966).

With *A. maculicornis* providing only partially effective biological control, attempts were made to acquire again this new species of *Coccophagoides.* The import came in 1951 to 1952 from India and Pakistan. In 1957 explorer entomologist Paul DeBach, while searching for a parasite of the citrus red scale in the Near and Far East, rediscovered the *Coccophagoides* sp. on olive scale infesting apple in Pakistan. Shipments of material to California led to the successful reacquisition of the new species in the quarantine laboratory. It was subsequently named *Coccophagoides utilis* Doutt.

The reason *C. utilis* was so difficult to rear was that while females are produced in the normal way as endoparasites on olive scale, males are produced only as hyperparasitic ectoparasites on immature female parasites of their own species in the same scale. This process of autoparasitism (adelphoparasitism of males on females of the same species) is common in the group of genera to which *Coccophagoides* belongs.

To rear this parasite, and later to mass-culture it for biological control colonizations, careful adjustments of the culture must be arranged. This species is arrhenotokous; mating is required to produce females. Mated females must be used to attack healthy olive scales, some of these latter must then be separated, held until the immature parasite reaches the pupal stage, and then exposed to unmated females for the production of male progeny. The remainder of the scales parasitized by mated females are then held for emergence of female progeny. Males are eventually recovered from parasite pupae (their own cousins) contained in the dead host scales. Only part of the newly emerged females can be exposed to males for mating; the remainder must be kept unmated to produce more males. This unusual biology can cause complications in culture and establishment, particularly in a program where there is an attempt to rear the thousands of parasites needed for colonization work.

After a culture technique for *C. utilis* was devised, the parasite was then colonized at several locations in California during the period 1957 to 1958 Establishment was proven to have occurred in 1961, when the parasite was recovered from two of the original release sites. It was found to occur in association with the previously established *A. maculicornis.* Subsequent studies disclose that rather than competing with each other to the detriment of the combined control effect, these two parasites supplemented one another. This supplementing effect was of such a nature as to result in excellent biological control of the scale, far better than either species alone could achieve, and as good as the control provided by the "best" insecticide.

While *A. maculicornis* was shown to be adversely affected by summer heat and low humidity, *C. utilis* on the other hand manages to survive the summers without undue losses. Being an excellent searcher of scales, it manages to produce levels of parasitization reaching 60 percent, even at low scale densities. So, while *A. maculicornis* thrives during the spring, though suffering in summers, *C. utilis* persists with little mortality during summer and supplements *A. maculicornis* during the spring.

This project thus provides interesting lessons, namely: (1) that it is important to discover the proper intraspecific natural-enemy strains, and (2) that two parasites can be better than one, that multiple introductions can definitely lead to improved results over single species introductions, and that potentially competitive parasites can coexist, provided external factors such as climate shift the competitive advantage from one to the other and back again.

KLAMATH WEED

This has been one of the most successful cases of weed control through use of plant-feeding insects, and it provides a very good example of how such a project should be prosecuted. It is also a very striking example of how important an herbivorous insect can be in influencing the ecology of a plant.

Klamath weed (*Hypericum perforatum*), also known as St. Johnswort, a perennial native of Europe and Asia, has invaded many semiarid to subhumid temperature regions of the world, notably Australia, New Zealand, Argentina, Chile, South Africa, and North America. It was first found in northern California, near the Klamath River (hence its American name) in the early 1900s. During the subsequent years, it spread throughout the low-elevation valleys, canyons, and coastal plains of California, occupying valuable rangelands, until by the mid 1940s it covered more than 2 million acres (Holloway and Huffaker 1952).

The weed is very aggressive in the dry western United States, and as it expanded its distribution, it displaced native and introduced forage plants of prime value to livestock. Besides displacing more suitable forage plants, Klamath weed is also poisonous to livestock. While chemical and cultural controls were possible as means of combating the weed, such measures were far too costly, the weed covered too great an area, and much of it was inaccessible.

In Australia a biological control program against *H. perforatum* was begun in 1920, with first attempts to obtain and test insect feeders of the weed concentrated in England. Lack of success caused a shift in search to southern France in 1935. There, three species of insects were found attacking the plant and exhaustive testing to determine the host specificity of the insects was commenced.

One of the major tasks in a biological control of weeds project is the determination, beyond any reasonable scientific doubt, that any candidate insect to be used in such a control program will not shift its attention from the target weed to some other plant, particularly one of economic importance such as a crop plant. Experience with insect feeding habits indicates that many phytophagous insect species are polyphagous, while many others are oligophagous, attacking several closely related plants. For weed control purposes an insect must be monophagous, or at least confine its attacks to closely related plant species which are themselves of no economic importance.

Two species of beetles from France, having passed the rigorous feeding tests beforehand, were sent to Australia and colonized in the late 1930s. Eight years later, observations showed the beetles were established and control results were promising.

In 1944 a similar biological control project was begun in California. Because of World War II, which prevented entomological work in France, the three beetles, *Chrysolina hyperici* (Forster), *C. quadrigemina* Suffrian, and *Agrilus hyperici* (Creutzer) were obtained from Australia and, after some additional host-specificity testing, were colonized in California in 1945 and 1946.

The beetles have but one generation per year, so that the results of a colonization could not be evaluated very rapidly. Nevertheless, within three years it was found that two of the leaf-beetle species (*Chrysolina* spp.) not only had become established, but were flourishing. The target weed was being severely defoliated at the release sites, so much so that within three years the hardy perennials were being killed back. (See Figure 14.)

The beetles spread slowly as long as Klamath weed was abundant, but later they were found to be capable of flights of as much as two to three miles radially from a point source. But by artificial distributions of field-collected beetles the entire infested area in California was colonized by 1950. Other western states also acquired the beetles.

The final results, attained by 1956 and still prevailing today, were striking. The dense, extensive stands of the weed were devastated. Only in certain shady sites in narrow canyons and under trees was the weed able to hold out, and even here it declined. Cattle growers were enthusiastic. Native

Figure 14. Biological control of Klamath weed. A, rangeland in Humboldt County, California partially cleared of the weed three years after beetle introduction. Note heavy Klamath weed stand in foreground; B, same location a year later with weed completely destroyed; C, cleared area repopulated with valuable forage plants. Photographs A and B by J. K. Holloway. Photograph C by J. Hamai.

forage plants (bunchgrass) and introduced clovers reoccupied the range (Huffaker and Kennett 1959). The control program was a resounding success, so much so that the grateful growers erected a monument to the successful beetles.

Ironically, the results of the control campaign against *H. perforatum* in Australia were nowhere as decisive. While suppression of the weed by the two *Chrysolina* spp. plus several other insects imported into Australia was noticeable, it was not as dramatic nor as complete as in the western United States (Huffaker 1966).

Attempts were made in 1951 to 1952 to establish the two *Chrysolina* species on Klamath weed in British Columbia, Canada (Smith 1958). Colonization attempts gave variable results, with permanent establishment occurring in only a few sites. Curiously, probably as a consequence of climatic stresses, it was not *C. quadrigemina* but *C. hyperici* that was the better performer, so far as relative numbers go. However, weed suppression was disappointing and biological control of *H. perforatum* in British Columbia remains to be accomplished.

The California experience provides evidence as to the stability of the plant-herbivore relation in insects used in weed control work. One of the major objections to biological control of weeds by use of phytophagous organisms raised by agricultural scientists and public authorities is the fear that insects, imported to control a plant by feeding on it, will shift their attentions to other plants. It was felt by these objectors that this would be particularly likely to occur as the employed insects accomplish a substantial destruction of the target pest plant, and then come under severe starvation pressures. However, the results of the very stringent feeding tests conducted on the candidate beetles by researchers before colonization, reinforced by the total lack of evidence since the project attained its success in the mid-1950s that plants other than *H. perforatum* were attacked and damaged in any way, indicates that the biological control of weeds, if properly conducted, is an entirely safe and effective procedure.

REVIEW AND RESEARCH QUESTIONS

1. What group of insect pests has biological control been most successful against?

2. Early attempts to control the walnut aphid by use of an imported parasitoid were unsuccessful. What feature of the failed parasitoid was found to be the key to later success?

3. What North American pest insect was controlled by a parasite not notably effective in its native land?

4. Discuss the differences and similarities in the two biological control programs against the oriental fruit fly and the olive scale.

5. The program against the Klamath weed provides an example of the importance of an herbivorous insect in limiting the abundance of a plant. Discuss this aspect of biological control, including in your discussion the matter of host specificity and the natural control of plants by animals.

BIBLIOGRAPHY

Literature cited

Bailey, L. H., and E. Z. Bailey. 1949. *Hortus second.* New York: Macmillan, 778 pp.

Blunck, H. 1957. *Pieris rapae* (L.) its parasites and predators in Canada and United States. *J. Econ. Ent.* 50: 835-836.

Boese, G. 1936. Der Einfluss tierscher Parasiten auf den Organismus der Insecken. *Z. Parasitenk.* 8: 253-284.

Broodryk, S. W., and R. L. Doutt. 1966. Studies of two parasites of olive scale, *Parlatoria oleae* (Colvée) in California. 2. The biology of *Coccophagoides utilis* Doutt (Hymenoptera, Aphelinidae). *Hilgardia* 37: 233-254.

Clausen, C. P., D. W. Clancy, and Q. C. Chock. 1965. *Biological control of the Oriental fruit fly* (*Dacus dorsalis* Hendel) *and other fruit flies in Hawaii.* U.S. Dept. Agric. Tech. Bull. 1322. 102 pp.

Davis, C. J. 1967. Progress in the biological control of the southern green stink bug *Nezara viridula* variety *smaragdula* (Fabricius) in Hawaii (Heteroptera: Pentatomidae). *Mushi* 39: 9-16.

DeBach, P. 1964. Successes, trends and future possibilities. In *Biological control of insect pests and weeds,* P. DeBach, ed. Chap. 24, pp. 673-713. London: Chapman & Hall, 844 pp.

DeBach, P., and L. C. Argyriou. 1967. The colonization and success in Greece of some imported *Aphytis* spp. (Hym. Aphelinidae) parasitic on citrus scales (Hom. Diaspididae). *Entomophaga* 12: 325-342.

Embree, D. G. 1971. The biological control of the winter moth in eastern Canada by introduced parasites. In *Biological control,* ed. C. B. Huffaker, Chap. 9, pp. 217-268. New York: Plenum Press, 511 pp.

Hafez, M., and R. L. Doutt. 1954. Biological evidence of sibling species in *Aphytis maculicornis* (Masi) (Hymenoptera, Aphelinidae). *Can. Ent.* 86: 90-96.

Hawkes, R. B. 1968. The cinnabar moth, *Tyria jacobaeae,* for control of tansy ragwort. *J. Econ. Ent.* 61: 499-501.

Holloway, J. K., and C. B. Huffaker. 1952. Insects to control a weed. In *Insects,* Yearbook of Agriculture for 1952, pp. 135-140. U.S. Govt. Printing Office, Washington, D.C.

Huffaker, C. B. 1966. A comparison of the status of biological control of St. Johnswort in California and Australia. In *Natural enemies in the Pacific area.* Eleventh Pac. Sci. Cong., Tokyo.

Huffaker, C. B., and C. E. Kennett. 1959. A ten-year study of vegetational changes associated with biological control of Klamath weed. *J. Range Mgt.* 12: 69-82.

Huffaker, C. B., and C. E. Kennett. 1966. Studies of two parasites of olive scale, *Parlatoria oleae* (Colvée). 4. Biological control of *Parlatoria oleae* (Colvée) through the compensatory action of two introduced parasites. *Hilgardia* 37: 283-335.

Huffaker, C. B., C. E. Kennett, and G. L. Finney. 1962. Biological control of olive scale, *Parlatoria oleae* (Colvée), in California by imported *Aphytis maculicornis* (Masi) (Hymenoptera, Aphelinidae). *Hilgardia* 32: 541-636.

Kobakhidze, D. N. 1965. Some results and prospects of the utilization of beneficial entomophagous insects in the control of insects in Georgian S.S.R. (USSR). *Entomophaga* 10: 323-330.

Mathys, G., and E. Guignard. 1965. Etude de l'efficacité de *Prospaltella perniciosi* Tow. en Suisse parasite du pou de San Jose. *Entomophaga* 10: 193-220.

National Academy of Sciences. 1968. The biological control of weeds. In *Principles of plant and animal pest control,* Vol. II, Chap. 6, pp. 86-119. Nat. Acad. Sci. Publ. 1597.

Parker, F. D. 1971. Management of pest populations by manipulating densities of both hosts and parasites through periodic releases. In *Biological control,* ed. C. B. Huffaker, Chap. 16, pp. 365-376. New York: Plenum Press.

Parker, F. D., F. R. Lawson, and R. E. Pennell. 1971. Suppression of *Pieris rapae* using a new control system: mass releases of both the pest and its parasites. *J. Econ. Ent.* 64: 721-735.

Puttler, B., F. D. Parker, R. E. Pennell, and S. E. Thewke. 1970. Introduction of *Apanteles rubecula* into the United States as a parasite of the imported cabbageworm. *J. Econ. Ent.* 63: 304-305.

Richards, O. W. 1940. The biology of the small white butterfly (*Pieris rapae*), with special reference to the factors controlling its abundance. *J. Animal Ecol.* 9: 243-288.

Shuster, M. F., J. C. Boling, and J. J. Marony, Jr. 1971. Biological control of Rhodesgrass scale by airplane releases of an introduced parasite of limited dispersing ability. In *Biological control,* ed. C. B. Huffaker, Chap. 10, pp. 227-250. New York: Plenum Press.

Smith, J. M. 1958. Biological control of Klamath weed, *Hypericum perforatum* L. in British Columbia. *Proc. 10th Int. Congress Ent. Montreal* (1956) 4: 561-565.

Tooke, F. G. C. 1953. *The eucalpytus snout-beetle,* Gonipterus scutellatus *Gyll. A study of its ecology and control by biological means.* Union S. Africa, Dept. Agric. Entomol. Mem. 3, 283 pp.

van den Bosch, R. 1971. Biological control of insects. *Ann. Rev. Ecol. and Systematics* 2: 45-66.

van den Bosch, R., E. I. Schlinger, and K. S. Hagen. 1962. Initial field observations on *Trioxys pallidus* (Haliday), a recently introduced parasite of the walnut aphid. *J. Econ. Entomol.* 58: 857-862.

van den Bosch, R., E. I. Schlinger, J. C. Hall, and B. Puttler. 1964. Studies on succession, distribution, and phenology of imported parasites of *Therioaphis trifolii* (Monell) in Southern California. *Ecology* 45: 602-621.

van den Bosch, R., B. D. Frazer, C. S. Davis, P. S. Messenger, and R. Hom. 1970. *Trioxys pallidus*—an effective new walnut aphid parasite from Iran. *Calif. Agric.* 24:8-10.

Wilkinson, A. T. S. 1966. *Apanteles rubecula* Marsh. and other parasites of *Pieris rapae* in British Columbia. *J. Econ. Ent.* 59: 1012-1018.

Additional references

Clausen, C. P. 1956. *Biological control of insect pests in the continental United States.* USDA Tech. Bull. 1139, 151 pp.

Huffaker, C. B., ed. 1971. *Biological control.* New York: Plenum Press, 511 pp.

Turnbull, A. L., and D. A. Chant. 1961. The practice and theory of biological control of insects in Canada. *Canad. J. Zool.* 39: 697-753.

Wilson, F. 1960. *A review of biological control of insects and weeds in Australia and Australian New Guinea.* Tech. Comm. No. 1, Commw. Inst. of Biol. Contr., 102 pp.

8 Naturally Occurring Biological Control and Integrated Control

Faunistic surveys of agricultural, sylvan, or natural, undisturbed environments will disclose large numbers of insect species which are of insignificant abundance, causing little or no harm to the plants growing in such habitats. Many of these insect species, phytophagous in habit, are kept in check by native natural enemies. We describe this situation as *naturally occurring biological control.*

Such natural enemies as those described above do not figure in conventional biological control efforts by man. In fact, because they and their hosts occur in low numbers, they are not often observed in the environments concerned. Such host species are rarely evaluated as pests and hence have usually been ignored by agricultural growers.

In the years before 1945, when pest controls (mainly chemicals) were used, their relative specificity or lack of persistence caused them but rarely to interfere with or upset this naturally occurring biological control. Furthermore, other methods of pest control, such as classical biological control, host plant resistance, and cultural control (see Chapter 9), were often used as solutions to pest problems.

Then in the years after 1945, when the powerful, synthetic organic insecticides appeared, they made insect suppression so easy and effective that virtually every other method of pest control was dropped in favor of these miracle chemicals as our insect control "crutch." But in neglecting the investigation and development of alternative controls we rendered ourselves vulnerable to insect depredation in the event that the chemicals failed. Very few imagined that this would ever come about. But today it is happening and because we do not have pest control systems which embrace alternative methods, major disaster threatens in many areas. The signs are manifold: In northeastern Mexico, the cotton industry has been destroyed by the tobacco budworm, an insect that cannot be killed by insecticides (Adkisson 1971); in Central America, the human population, virtually malaria-free for more than a decade, is now threatened with a devastating epidemic as the vector mosquito, *Anopheles albimanus* Wiedemann, verges on total resistance to available insecticides (Georghiou 1971); in California's Great Central Valley similar resistance has developed in the encephalitis vector, *Culex tarsalis* Coquillet (Georghiou 1971); the world over, in the wake of insecticide usage, spider mites have leapt into the forefront as crop pests and threaten economic disaster in a variety of commodities (Huffaker et al. 1970; McMurtry et al. 1970). Today there are more insect species of pest status than ever before, insect control costs have mounted strikingly, and insecticide pollution has become a problem of global proportions.

EFFECTS OF INSECTICIDE

A number of factors have contributed to this alarming situation, but insecticide disruption of naturally occurring biological control lies very much at the heart of the matter. In this connection, it is now glaringly apparent that in the mid 1940s, when DDT burst onto the scene, there was an appalling lack of awareness of the role of naturally occurring biological control in insect population regulation. Quite evidently, some entomologists knew that natural enemies could control insect populations for, as described here, a number of exotic pests had been controlled by introduced natural enemies up to 1940. But it seems that hardly anyone gave thought to the possibility that the same process of population regulation could affect endemic pests; that native species might be held at innocuous levels by natural enemies. In other words, a sort of narrow-minded view appears to have existed concerning the role of natural enemies in pest control; on the one hand, the deliberate reassociation of invading insects and their natural enemies (classical biological control) was accepted by many as a logical pest control tactic, but the inverse of this, the possibility that the disassociation (e.g., as by broad spectrum insecticides) of natural enemies from their native hosts might permit the explosion of the latter to severe pest status seems not to have occurred to very many persons.

One is forced to this conclusion by the events which took place at the time of the DDT breakthrough and in the years thereafter. Entomological history clearly reveals that at the outset of the synthetic organic insecticide era (i.e., the late 1940s), the great majority of those involved in insect control plunged ahead with the new chemical tools essentially oblivious to the ecological (and genetic) pitfalls that lay ahead. There simply was no general realization that the materials had inherent faults which predetermined the development of serious problems. Even the early signs of impending trouble—unprecedented outbreaks of spider mites, scale insects, and various lepidopterous larvae following the use of DDT, and the rapid development of resistance to that material in the housefly and certain mosquito species—failed to create widespread recognition of what was impending. These developments were simply viewed by most entomologists as minor aberrations that could be corrected by the substitution or addition of other chemicals. Now, a quarter of a century later, the problems have become too varied, widespread, and serious to be further ignored. It is quite apparent that the very characteristics that made the new insecticides effective insect-killers have led to the troubles arising from their use (Huffaker 1971).

The basic flaw in the modern insecticide is its broad toxicity. The materials kill indiscriminately, and when applied in the field their broadly toxic action can virtually strip the treated areas of arthropod life. Thus, where they are used, the ecosystemic web is often shattered overnight, and a biotic vacuum created in which violent reactions are almost inevitable. The overall problems would not have been so serious if the development and exploitation of the synthetic organic insecticides had occurred as an orderly process. But this was not the case. The spectacular performance of DDT had a catalytic effect on the insecticide industry so that an outpouring of other organochlorine and of organophosphate materials quickly followed. Thus, in a span of only five or six years following World War II, a wide spectrum of synthetic organic materials was made available. The enthusiasm of researchers and users, coupled with the highly effective merchandising apparatus of the agrochemical industry gave great impetus to the widespread field application of these materials.

Ecological backlash

As some had foreseen, inevitable backlashes quickly occurred literally everywhere that the materials were used. The agriculturalist placed himself on an insecticidal treadmill because of three ecologically based phenomena: (1) target pest resurgence, (2) secondary pest outbreaks, and (3) pesticide resistance. All of these are directly related to the disruption of biological control and lead increasingly to a "pesticide addiction" from which it is difficult to withdraw (Stern et al. 1959, Smith and van den Bosch 1967). This concept is

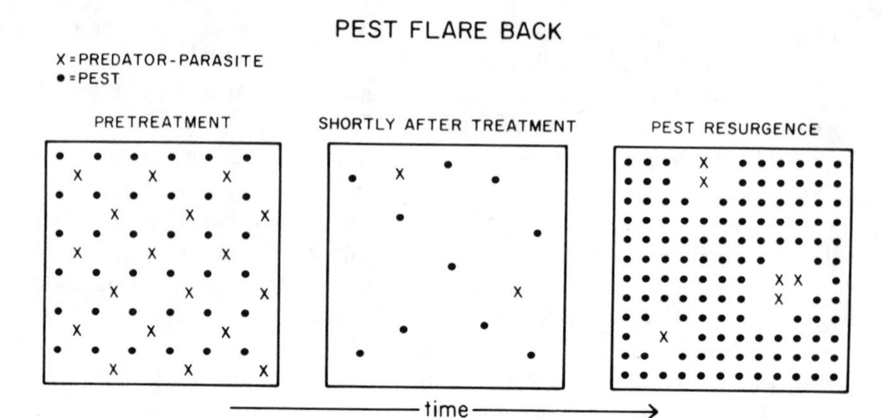

Figure 15. Target pest resurgence. Diagrammatic sketch of the influence of chemical treatment on natural-enemy pest abundance and dispersion with resulting pest resurgence. The squares represent a field or orchard immediately before, immediately after, and some time after treatment with an insecticide for control of a pest species represented by the solid dots. The immediate effect of treatment is a strong reduction of the pest, but an even greater destruction of its natural enemy (enemies), represented by X's. The resulting unfavorable ratio and dispersion of hosts (pest individuals) to natural enemies permits a rapid resurgence of the former to damaging abundance. From Smith and van den Bosch 1967, with permission of the publisher.

diagrammed in Figures 15 to 19. Thus, target pest resurgence occurs when the insecticide not only kills high percentage of the pest population but also destroys a large proportion of its natural enemies, permitting the rapid return and eventual population explosion of the pest. (See Figures 15 and 16.) Secondary pest outbreaks occur where insecticides applied to control noxious insects destroy the natural enemies of innocuous species occupying the same habitat; the latter, freed of their biological controls, then erupt to damaging levels. (See Figures 17 and 18.) Target pest resurgence and secondary pest outbreaks in turn contribute directly to insecticide resistance because they necessitate heavier and more frequent treatments which speed genetic selection for resistance. (See Figure 19.)

Economic problems

The seriousness of the trifaceted insecticide treadmill and its consequence, "pesticide addiction," cannot be exaggerated. It is global in extent and has

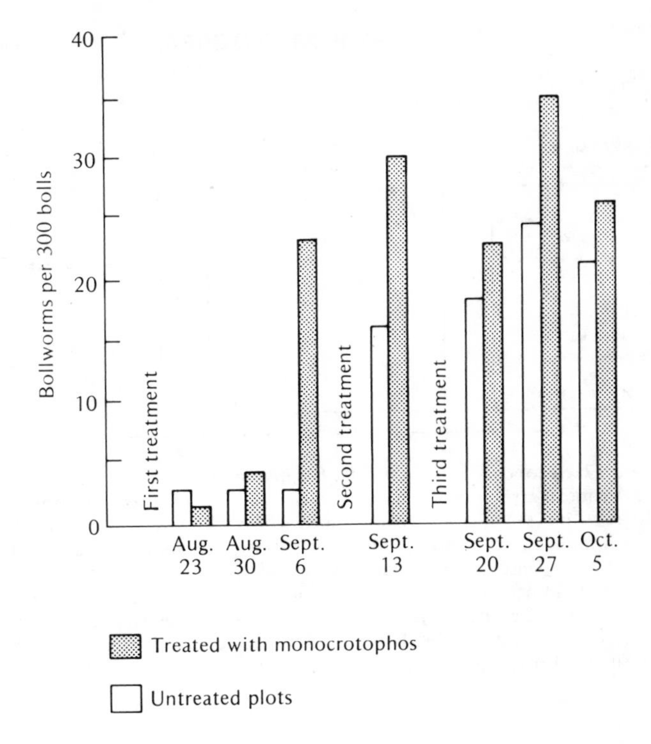

Figure 16. Target pest resurgence following applications of a "control" insecticide. In this experiment, plots treated with monocrotophos, an insecticide federally registered for bollworm control, suffered heavier infestations than untreated plots. Simultaneous samplings of predators revealed that the insecticide destroyed bollworm predators which permitted resurgence of the pest. The data are from an experiment conducted at Dos Palos, California, in 1965.

contributed massively to economic and ecological problems virtually everywhere that modern insecticides have been used (Huffaker 1971). Nothing illustrates this more clearly than the universal spider mite problem. A quarter of a century ago, spider mites (Tetranychidae) as a group were minor pests. Today they are the most serious arthropod pests affecting agriculture worldwide (Huffaker et al. 1970, McMurtry et al. 1970). For example in California alone spider mites annually cost agriculture about $60 million, or about one-fifth of all the costs and losses attributable to pest arthropods. A single species, the citrus red mite, *Panonychus citri* (McGregor), a minor pest

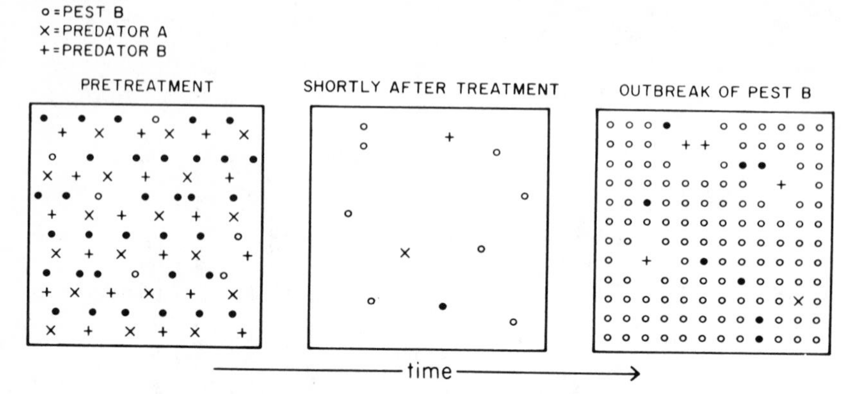

Figure 17. Secondary pest outbreak. Diagrammatic sketch of the influence of a chemical treatment on natural-enemy pest abundance and dispersion with resulting secondary pest outbreak. The squares represent a field or orchard immediately before, immediately after, and some time after treatment with an insecticide for control of pest A represented by (●). The chemical treatment effectively reduces pest A as well as its natural enemy (X), but has little or no effect on pest B (○). Subsequently, because of its release from predation, pest B flares to damaging abundance. From Smith and van den Bosch 1967, with permission of the publisher.

twenty-five years ago, now costs the citrus industry more than $10 million annually, which is about one-half the total cost of pest arthropods in citrus (Hawthorne 1970).

A wide variety of arthropods prey heavily on spider mites, with predatory mites of the family Phytoseiidae perhaps being the most important group, although tiny lady beetles (*Stethorus* spp.) are also acclaimed by many. Species in several other families of the Acarina also prey on tetranychids. In the Insecta, important spider mite enemies occur in the Coleoptera (Coccinellidae, Staphylinidae), the Neuroptera (Chrysopidae, Hemirobiidae, Coniopterygidae), Hemiptera (Anthocoridae, Miridae, Nabidae, Lygaedae), and Thysanoptera (Scolothrips) (Clausen 1940, Huffaker et al. 1970, McMurtry et al. 1970). Individually and collectively these predators often effect substantial to complete control of spider mites, and it has largely been the disruption of this naturally occurring biological control which has triggered the global problem.

The worldwide spider mite problem and its causes have recently been thoroughly analyzed (Huffaker et al. 1970, McMurtry et al. 1970). These reviews present overwhelming evidence that the globally aggravated spider mite problem is largely a result of disruption by insecticides of the biological controls affecting these pests. In other words, it is the familiar triad of target pest resurgence, secondary pest outbreak, and pesticide resistance. Some spider mites are now so resistant to control chemicals that no available material gives adequate relief.

The arthropod pest problems plaguing cotton the world over illustrate equally well the vast importance of naturally occurring biological control. Wherever cotton is grown, it is affected by a variety of arthropod pests, and since it is a cash crop or foreign exchange earner, the first inclination is to

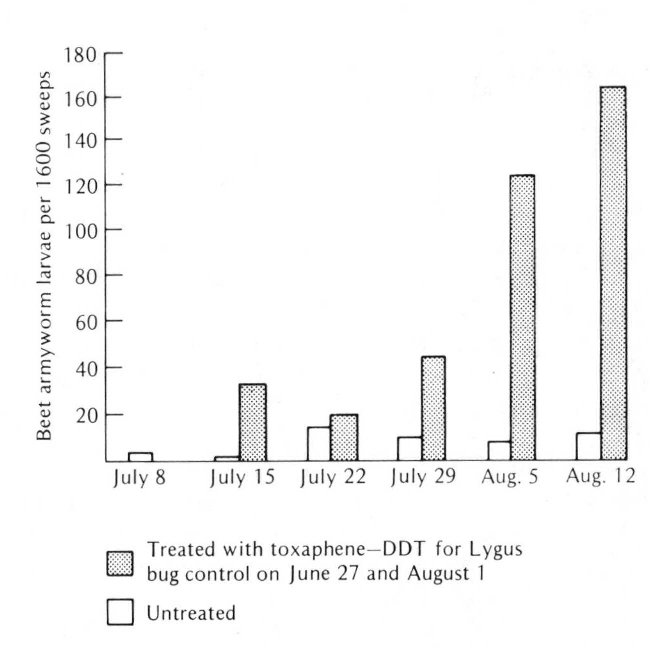

Figure 18. A secondary outbreak of the beet armyworm in cotton following insecticide treatment. Note that on August 12 there were approximately seventeen times as many armyworms in plots that had been previously treated with insecticide (a toxaphene-DDT mixture) as in plots with no history of treatment. Data are from an experiment conducted near Corcoran, California, in 1969. Subsequent studies have clearly shown that elimination of predators in the treated cotton permits such secondary pest explosion.

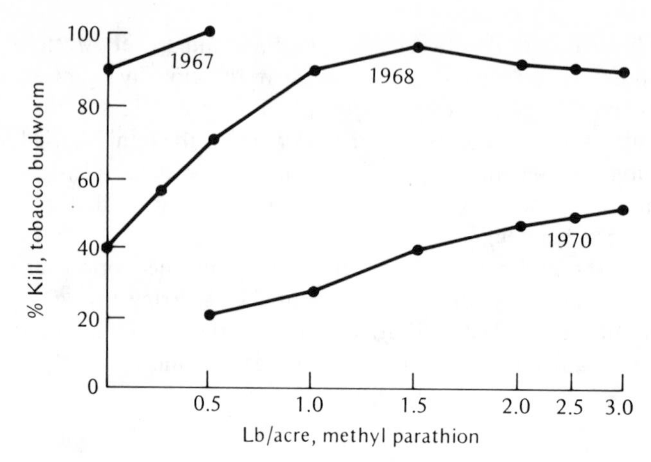

Figure 19. Insecticidal resistance. Decline in toxicity of methyl parathion to tobacco budworms from central Texas during the period 1967-1970. Data from P. L. Adkisson, Texas A. & M. University, College Station.

protect this source of income from pest depredation. As a consequence, cotton growers are perhaps the world's greatest users of insecticides, and they have paid a severe price for this heavy pesticide load. From virtually every area on earth where cotton is grown reports have come in recent years of economic, ecological, and even sociological problems directly attributable to insecticides (Adkisson 1971, Smith and Reynolds 1972, van den Bosch 1971). And again, as with spider mites, the trouble largely results from the disruption of naturally occurring biological control. For example, in northeastern Mexico, the tobacco budworm, *Heliothis virescens* (Fabricius), unleashed from its natural enemies by insecticides applied for control of early season pests (e.g., boll weevil, plant bugs), rapidly developed resistance to all insecticides and within a decade totally destroyed a $50 million cotton industry. When the cotton industry was moved to a new area the tobacco budworm destroyed it there in only four years (Adkisson 1971). In California's Imperial Valley, insecticides applied for control of the pink bollworm destroyed the parasites of the cotton leaf perforator, *Bucculatrix thurberiella* Busck, and this pest, which is resistant to virtually all insecticides used in cotton, then erupted to enormous abundance and defoliated thousands of acres (Reynolds 1971). Other areas where particularly severe insecticide associated problems have developed in cotton include the Cañete Valley of Peru, Central America, the Rio Grande Valley of Texas, and in Colombia, Egypt, and Turkey (Smith and Reynolds 1972).

In each of these places target pest resurgence, secondary pest outbreaks, and insecticide resistance have characterized the problem, and again these have largely resulted from the interference of the insecticides with naturally occurring biological control.

The spider mite and cotton pest problems only partially depict the worldwide biotic backlash to modern insecticide usage, for similar developments have occurred in many other pest control situations (Huffaker 1971). However, space does not permit a broader treatment of the matter here. Instead, it is hoped that the preceding discussion has provided insight into the scope and magnitude of naturally occurring biological control and the critical need to take this great natural force into account in future insect control schemes.

INTEGRATED CONTROL

The expanding problems generated by the adverse effects of the modern insecticides on natural enemies have led not only to an unprecedented appreciation of naturally occurring biological control, but also to an increasing realization that a philosophy of insect control which is dominated by a single tactic is programmed for failure. Consequently, a new and more sophisticated insect control philosophy has emerged, and programs based on this philosophy are coming into increasing application. The newly emergent concept has been termed *integrated control* (Stern et al. 1959, Smith and van den Bosch 1967).

Integrated control is a pest population management system that utilizes all suitable techniques (and information) either to reduce pest populations and maintain them at levels below those causing economic injury or to so manipulate the populations that they are prevented from causing such injury. Integrated control achieves this ideal by harmonizing techniques in an organized way, by making the techniques compatible, and by blending them into a multifaceted flexible system (Smith and Reynolds 1966). Its goal is not simply pest annihilation but, what is more important, the reduction of pest populations only to levels compatible with the economic production of the crop, and the concurrent maintenance of environmental integrity. In other words, the ultimate goal of any integrated control program is the economical and ecologically acceptable management of pest insect populations.

It is the emphasis on the fullest practical utilization of the existing regulating and limiting factors in the ecosystem which gives integrated control its uniqueness. Perhaps the greatest advantage of the approach is that in maximizing the role of naturally occurring mortality and regulating factors, it automatically assures a high level of environmental quality, since such maximization can only be attained where there is minimum disruption of the environment. A second major advantage is economy, which again derives from its heavy reliance on natural controls and minimal dependence on costly artificial measures.

There are two basic considerations in the development of an integrated control program: (1) the ecosystem, and (2) valid economic thresholds for the pest species, the latter based on accurate assessments of the pest's damaging potential and the ecological, sociological, and external economic costs created by any artificial means (e.g., insecticides) used for control.

By contrast, conventional insect control centers on maximum destruction of the pest species, and except for consideration of the hazard to man, domesticated animals, some wildlife (particularly sport fish and game), and perhaps the honeybee, other elements in the ecosystem are ignored. Economic justification for use of insecticides is almost invariably poorly established or not established at all. Only too frequently, the essential criterion for pesticide use is the highly simplistic one of mere pest presence, and the hallmark of pesticide efficiency is maximum kill.

In initiating an integrated control program the automatic first step is a consideration of the ecosystem, particularly as regards the possible effects of any artificial control practices on it, and an assessment of the factors in the ecosystem which favor or militate against populations of pest or potential pest species. The second step is the critical analysis of the "pest" status of the reputedly injurious species. The integrated control philosophy rejects the use of insecticides simply because pest insects occur in an area or threaten to occur there. Instead, under integrated control, criteria are developed to pinpoint those times and places where insecticides are truly needed. It follows that where such need exists, every effort is made to effectively use the safest, most ecologically tenable materials.

To help clarify the differences between conventional insect control and integrated control, Table 5 compares the two types of programs for cotton insect control in California's San Joaquin Valley. The development of the integrated control program in San Joaquin Valley cotton has embraced a wide range of studies in a highly coordinated program: analysis of cotton growth characteristics and fruiting patterns relative to insect phenologies, feeding habits, and population levels; investigations of insect host affinities; analyses of distribution, migration, and phenology of *Lygus hesperus*; studies on the phenologies and in-field and on-plant distribution of *H. zea* and the other lepidopterous species; analyses of the natural-enemy complexes affecting the

Table 5. Comparison of conventional and integrated pest control programs in cotton in California's San Joaquin Valley.

Pest	*Conventional Control*	*Integrated Control*
Lygus hesperus	Treatment threshold	Treatment threshold
	10 bugs/50 net sweeps at any time of season from about June 1 until about mid-August, or prophylactic applications based on time of season, grower apprehension, sales advice, etc.	When population levels of 10 bugs/50 net sweeps are recorded on two consecutive sampling dates (3- to 5-day interval) during flower budding period (approximately June 1 to mid-July). Treatments after mid-July are discouraged as being essentially useless and highly disruptive to natural enemies of lepidopterous pests.
	Cultural controls	Cultural controls (in limited use)
	None.	(a) Strip harvesting of alfalfa hay fields to prevent migration of bugs into adjacent cotton when alfalfa is mowed. (b) Interplanting of alfalfa strips in cotton fields to attract *Lygus* out of the cotton. Both practices are based on the strong preference of *L. hesperus* for alfalfa over cotton.
Lepidopterous Pests The complex of lepidopterous pests in San Joaquin Valley cotton includes three major species: *Heliothis zea* (bollworm), *Trichoplusia ni* (cabbage looper), and *Spodoptera exigua* (beet armyworm). All three species are normally under heavy pressure from natural enemies and usually only erupt to great abundance in the wake of insecticide treatments.		

Table 5. (Continued)

Pest	Conventional Control	Integrated Control
Heliothis zea	Treatment threshold (a) 4 small (¼" or less in . length) larvae/100 inspected plants. This is a long-standing treatment level of unknown origin which has been found to be completely erroneous. Many fields in the San Joaquin Valley sustain infestations of this magnitude and consequently there is much unnecessary treatment for bollworm control. (b) Nebulous criteria such as time of season, grower apprehension, sales pressure, etc. are also involved in insecticide treatment decisions for bollworm control.	Treatment threshold (a) 15 small larvae/100 inspected plants where field has been previously treated with insecticides. (b) 20 small larvae/100 inspected plants, where fields have not been previously treated with insecticides. Infestations of this magnitude rarely develop in untreated cotton fields in the San Joaquin Valley. Consequently cotton that has reached the period of major *H. zea* activity (i.e., late July and early August) free of insecticide treatment is virtually unthreatened by this pest.
Trichoplusia ni	Treatment threshold There is no established economic threshold for this "pest," instead chemical control is invoked on the basis of such nebulous criteria as "when abundant," "when damage is evident," "when defoliation is threatened." The cabbage looper is commonly abundant in conventionally managed cotton fields where it erupts in the wake of early and midseason insecticide treatments, particularly those applied for *Lygus* control. Consequently, San Joaquin Valley cotton is heavily treated for its control. The insect is very difficult	Treatment threshold There is no evidence that this insect causes economically significant damage to cotton. It is rarely abundant in integrated control fields, since its populations are normally controlled by predators and parasites. Consequently, in those fields that have been in the integrated control program, there have been no insecticidal treatments for cabbage looper control.

Table 5. (Continued)

Pest	Conventional Control	Integrated Control
	to kill and as a result, heavy dosages of insecticide mixtures are used against it. Needless to say, such treatments add substantially to pest control costs and they increase the health hazard and environmental pollution.	
Spodoptera exigua	Treatment threshold There is no established treatment threshold for this species, and the same criteria are utilized in making decisions for its chemical control as with cabbage looper. However, ongoing research indicates that where abundant the insect can cause economically significant injury to cotton, but no specific treatment threshold has yet been developed. The beet armyworm is essentially a secondary outbreak pest, and like cabbage looper it is very difficult to "control." Consequently, in the conventionally managed fields it often receives costly, pollutive chemical treatments.	Treatment threshold The best armyworm has not been abundant in the integrated control fields, and there has been no need to consider its control.
Spider Mites (Tetranychidae)	Treatment threshold There are no established economic thresholds for spider mites in San Joaquin Valley cotton. A number of selective and nonselective acaricides are used prophylactically and therapeutically for spider mite "control." Again, as in cabbage looper and beet armyworm control the treatment criteria are largely nebulous.	Treatment threshold No established economic thresholds. Research on this matter is currently under way, as is research on selective chemical controls and the development of spider-mite-resistant cotton. In the fields that have been under the integrated control program the growers have at times used a selective acaricide prophylactically.

various pest species and the impact of these enemies on their hosts; studies on nutritional augmentation of natural enemies; determination of the effect of various chemical insecticides on natural enemies and the cotton plant itself; testing of chemical insecticides and microbial agents on the several pest species; investigations on timing of the various chemical and microbial control treatments; studies on the effect of the feeding of the various pests on cotton growth and fruiting characteristics; analysis of the effects of irrigation and fertilization on populations of *L. hesperus*; studies on a number of other factors in the cotton ecosystem.

These data have not been accumulated in a haphazard, disjointed way. Instead, many are forwarded as they are gathered to a processing team which subjects them to computer analysis and stores the computerized data in a central file for use by the entire research group. Furthermore, the group meets about twice a year to discuss the various aspects of the ongoing studies. During these meetings, decisions are reached concerning shifts in research emphasis, new lines of study to be undertaken, and adjustments in management recommendations for the various pest species.

The integrated control program for cotton insects in the San Joaquin Valley is not yet a finished product, but where it has been brought into its fullest application, insecticide usage has been reduced by more than 50 percent. This of course has brought economic benefit to the growers and has correspondingly reduced the pollution threat from insecticides.

Throughout the world, in addition to cotton, integrated control programs have been developed in alfalfa, apple, citrus, grape, peach, pear, oil palm, rubber, cocoa, and glasshouse crops (Huffaker 1971). In each of these cases preservation and/or augmentation of natural enemies has been a critical factor in the program. Currently, many additional programs are under development and these, too, place major emphasis on the role of naturally occurring biological control.

The pattern of today's research clearly reflects the expanding emphasis on integrated control and the associated recognition of the importance of naturally occurring biological control. In this light, it is quite apparent that as time passes the research effort focused on naturally occurring biological control and its utilization in pest insect control will almost surely equal or perhaps even outshadow the efforts in classical biological control. But whatever the case, it is highly unlikely that naturally occurring biological control will ever again be generally ignored in insect control considerations.

REVIEW AND RESEARCH QUESTIONS

1. Why are most native insects rare in numbers in our agricultural environments?

2. Discuss the relations of pesticides to biological control.

3. What is meant by the term *secondary pest outbreak*? Give an example.

4. What is integrated control?

5. Contrast conventional chemical control with integrated control.

BIBLIOGRAPHY

Literature cited

Adkisson, P. L. 1971. Objective uses of insecticides in agriculture. In *Agricultural chemicals—harmony or discord for food-people-environment,* ed. J. E. Swift, pp. 43-51. Univ. Calif. Div. Agric. Sci. 151 pp.

Clausen, C. P. 1940. *Entomophagous insects.* New York: McGraw-Hill, 688 pp.

Georghiou, G. P. 1971. Resistance of insects and mites to insecticides and acaricides and the future of pesticide chemicals. In *Agricultural chemicals—harmony or discord for food-people-environment,* ed. J. E. Swift, Univ. Calif. Div. Agric. Sci. 151 pp.

Hawthorne, R. M. 1969. *Estimated damage and crop loss caused by insect/mite pests—1968.* Calif. Dept. Agric. E-82-11. 11 pp.

Huffaker, C. B., ed. 1971. *Biological control.* New York: Plenum Press, 511 pp.

Huffaker, C. B., M. van de Vrie, and J. A. McMurtry. 1970. Ecology of tetranychid mites and their natural enemies: A review. II. Tetranychid populations and their possible control by predators. *Hilgardia* 40: 391-458.

McMurtry, J. A., C. B. Huffaker, and M. van de Vrie. 1970. Ecology of tetranychid mites and their natural enemies: A review. I. Tetranychid enemies. Their biological characters and the impact of spray practices. *Hilgardia* 40: 331-390.

Reynolds, H. T. 1971. A world review of the problem of insect population upsets and resurgences caused by pesticide chemicals. In *Agricultural chemicals—harmony or discord for food-people-environment,* ed. J. E. Swift, pp. 108-112. Univ. Calif. Div. Agric. Sci. 151 pp.

Smith, R. F., and H. T. Reynolds. 1966. Principles, definitions and scope of integrated pest control. *Proc. FAO Symp. on Integrated Pest Control* I: 11-17.

Smith, R. F., and H. T. Reynolds. 1972. Effects of manipulation of cotton agro-ecosystems on insect populations. In *The careless technology: ecology and international development,* ed. M. T. Farvar and J. P. Milton. New York: Natural History Press, 1030 pp.

Smith, R. F., and R. van den Bosch. 1967. Integrated control. In *Pest control: biological, physical and selected chemical methods,* ed. W. W. Kilgore and R. L. Doutt, Chap. 17, pp. 295-340. New York: Academic Press, 477 pp.

Stern, V. M., R. F. Smith, R. van den Bosch, and K. S. Hagen. 1959. The integration of chemical and biological control of the spotted alfalfa aphid. Part 1. The integrated control concept. *Hilgardia* 29: 81-101.

van den Bosch, R. 1971. The melancholy addiction of ol' king cotton. *Natural History* 80(10): 86-91.

9

Other Biological Methods of Pest Control

Besides the use of natural enemies, there are several other methods of a biological nature which have been applied with success in the suppression of noxious species. These methods are based on host plant (or animal) resistance to pest insects or diseases, cultural control, control through use of sterile insects, and genetic control.

In several cases, one or another of these methods, used alone, has resulted in the satisfactory suppression of a serious pest. In a few cases, combinations of two or more of the techniques have been used with or without pesticides, for control of a given pest, for example, the control of the spotted alfalfa aphid in California by use of parasites and resistant plant varieties. In these latter situations, any one method alone usually has not been sufficient to provide fully adequate control. Use of several methods at the same time is called *integrated control* (see Chapter 8).

HOST PLANT RESISTANCE

The use of crop varieties resistant to attack or damage by pests has been practiced in North America since the later decades of the nineteenth century (Painter 1951; Anon. 1969). Although the first effective application of this technique did not occur in North America, still it had its roots in this continent. This case involved the grape phylloxera, *Phylloxera vitifoliae* (Fitch), a North American root aphid that was accidentally introduced into Europe. The European grape, *Vitis vinifera* Linnaeus, was extremely susceptible to the phylloxera, whose attack was so devastating that European viticulture was threatened with destruction. The use of rootstocks of American grape, *Vitis labrusca* Fox, on which were grafted European table and wine grape varieties, provided a rapid and effective solution to this problem.

The grape phylloxera-resistant rootstock example is one where man discovered and took practical advantage of the natural occurrence of resistance. However, in many subsequent examples, man not only made use of resistance, he actually created resistant varieties of plants and animals by hybridizing and deliberate, artificial selection. This points up the basic requirement for the development of resistant varieties, which is that the tendency towards resistance to pest attack or damage must be an inherited or genetic trait.

Early examples of the development of plant resistance by selective breeding include the creation, by the early 1900s, of potato varieties resistant to potato late blight, *Phytophthora infestans* (Montemartini) DeBary, the disease responsible for the Irish famine of the mid-nineteenth century.

By the early 1920s, the selective breeding of wheat varieties led to the creation of new varieties resistant to the major insect pest of wheat in North America, the Hessian fly, *Mayetiola destructor* (Say).

Since these beginnings there has been a steady, albeit slow, progress in the development and use of resistant varieties until at the present time some of our principal, devastating insect and disease pests have been overcome by this method (Painter 1951, Beck 1965).

The method may be generally described as follows. In fields of the crop plant heavily infested with the pest in question, a search is conducted, sometimes over a wide geographical area, for individual plants seemingly free of or little affected by attack. These plants are removed to the plant breeding nursery or greenhouse, and seeds are produced and collected. Seedlings are grown, and crossed with each other to produce hybrid clones of plants. Tests are conducted to assure that resistance is retained. The clones are also studied to assure that other necessary plant characteristics are present, such as vigor, yield potential, and growth qualities. Where resistance is inherited and other qualities are retained, or are restored by additional hybridization with parental stocks, the new, selected variety is then provided for general use.

Careful observation of the nature of insect-resistance in crop plants shows that resistance can be ascribed to one or more of three types, (1) tolerance, (2) antibiosis, or (3) nonpreference (Painter 1951). *Tolerance* occurs when the resistant plant can sustain normal levels of pest infestation without damage or loss in yield. *Antibiosis* refers to the suppression of attack or damage through some physiological effect on the pest, such as reduction in size, vigor, or developmental rate of the growing pest, or reduction of fecundity of the adult female, or the decrease in survival rate of the pest during its life cycle. *Nonpreference* refers to the reduction of attack brought about by the unattractiveness of the host or the inability of the pest to recognize it.

In many closely observed cases of host plant resistance, any two or all three of these types of resistance mechanisms may occur together. Among the wheat varieties resistant to the Hessian fly, the midwestern variety Pawnee is tolerant of the pest. With other wheats, antibiosis and tolerance together are involved. An example of nonpreference concerns the resistance in corn varieties to grasshopper attack, and in certain *Solanum* species to attack by the Colorado potato beetle. The resistance of certain midwestern corn varieties to the European corn borer involves contributions from all three modes of resistance—tolerance and antibiosis in substantial amounts, the nonpreference to a lesser degree.

Other successes in the discovery and development of resistance in crop plants involve the wheat-stem sawfly in Canada and the United States; the creation of varieties of corn resistant to fall armyworm, stalk borers, cornleaf aphid, and corn earworm; and wheat, oat, and barley varieties resistant to the cereal leaf beetle and the greenbug.

In the case of the corn earworm, resistance in certain corn varieties was of only enough benefit to enable *a reduction* in the number of pesticide applications needed to control this pest. However, even this outcome can lead to considerable value in reducing the costs of corn production and the side effects of excessive pesticide usage.

One of the more successful cases of host plant resistance development in a large crop devastated by a new insect pest concerns the spotted alfalfa aphid, *Therioaphis trifolii.* This pest invaded North America for the first time about 1954, in the southwestern states of New Mexico, Arizona, and California. Within two to three years it had spread throughout thirty other states. The pest thrives on alfalfa, causing death of individual plants, loss of vigor and reduction of growth, and the fouling of cropped hay and harvest equipment by the excessive accumulations of honeydew (aphid excreta).

In several alfalfa breeding nurseries in southern California and Nevada, where work was proceeding on the development of varieties resistant to certain plant disease such as wilt and to stem nematodes, the spotted alfalfa aphid proliferated. However, among the disease-resistant varieties it was noted that several plants of the variety Lahontan showed resistance to the aphid. These were collected, bred, and crossed to produce a very effective spotted alfalfa aphid-resistant Lahontan, a variety now used to a considerable extent over much of the southwestern United States. Moapa is another alfalfa variety resistant to the spotted alfalfa aphid and used in much of the southwestern United States. It was derived, after further hybridizing and selection from a susceptible variety known as African.

The use of these two resistant varieties of alfalfa, coupled with the biological control of the pest aphid by imported parasites and native predators, has enabled the continued production of this important forage crop. In Kansas, however, neither Lahontan nor Moapa is well adapted to the local conditions. The variety most extensively grown there is called Buffalo. The successful search for and crossing of individual plants of the Buffalo variety found resistant to the spotted alfalfa aphid (see Figure 20) eventually led to the creation of a new, resistant variety called Cody. This was found to be as good as or even better than Lahontan when grown in Kansas conditions (Painter 1960).

In most situations only those growers who adopt resistant varieties of crop plants benefit from the practice, while those who retain the susceptible varieties continue to suffer damage (Painter 1960). However, in some cases, the use of resistant varieties produces "spill-over" benefits. Such is the case with wheat varieties resistant to the Hessian fly in Kansas. With the widespread adoption of resistant wheats, the overall incidence of Hessian fly declined by as much as 50 percent leading to reduction of damage, even in fields planted to nonresistant wheat.

Host plant resistance is by no means permanent, since the pest itself has the ability to adapt to the new varieties. However, the breakdown of resistance in initially resistant varieties is often a slow process, and with appropriate effort it can be counteracted by continuing search for additional resistant stocks. It is, on the other hand, relatively very inexpensive, once the proper teamwork between plant breeder, agronomist, and pest expert has been developed. Host plant resistance is sometimes the only practicable insect control method where crop production economics does not justify use of expensive pesticides.

Plant varieties resistant to diseases are also numerous. In fact, maize varieties resistant to leaf rust have been known since prehistoric times. Flax varieties have been developed (1908 to 1925) which are resistant to *Fusarium* wilt. Sugarcane varieties resistant to mosaic have been in use since 1926. Wheats resistant to wheat-stem rust, *Puccinia graminis* Persoon, although ephemeral in durability, have been in existence, with millions of bushels saved, since 1938.

Nematode resistance has also been successfully developed in crop plants including alfalfa, bean, barley, tobacco, potato, rice, and tomato.

CULTURAL CONTROL

Cultural control means the development or adjustment of agronomic or horticultural procedures so as to reduce pest abundance or minimize or

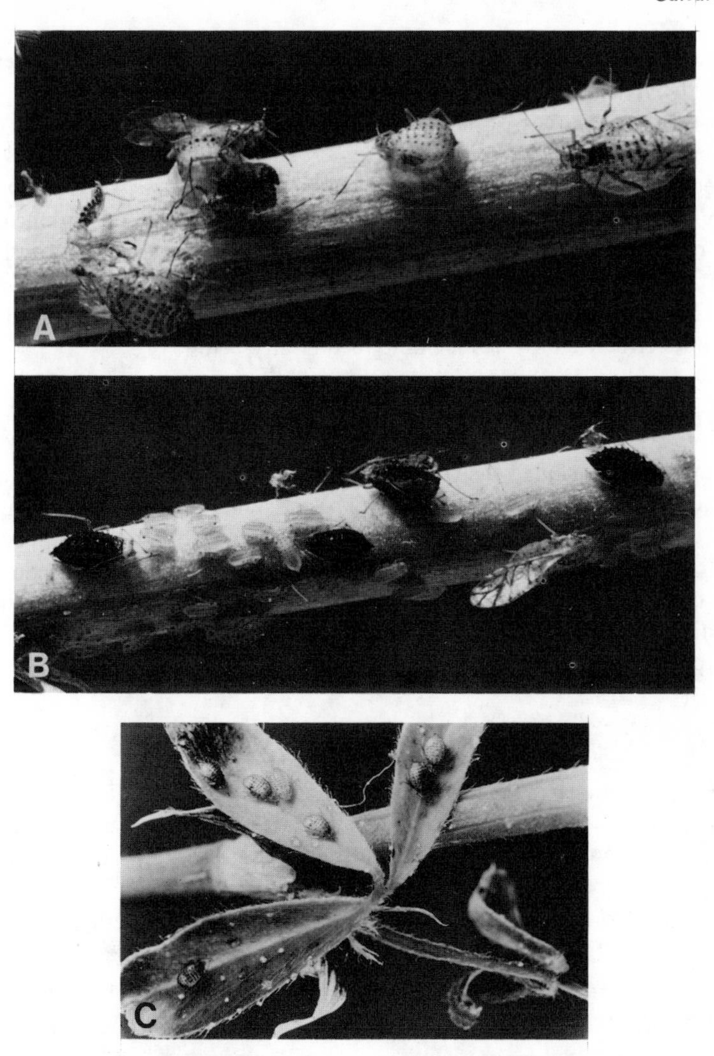

Figure 20. Parasitized spotted alfalfa aphids. A, aphids mummified by *Praon exsoletum.* Note the tentlike *Praon* cocoons beneath the mummified aphid skins; B, Mummies formed by *Aphelinus asychis* in a colony of living aphids. Note the striking black color of the mummified aphids; C, Aphids mummified by *Trioxys complanatus.* Note the slightly darker coloration and more turgid shape of the mummies as compared to the single living aphid in the center of the middle leaflet. Photographs by Ken Middleham, University of California, Riverside.

prevent damage resulting from pests (Anon. 1969). Stated another way it is the alteration of the agricultural environment so that, while still producing a suitable crop, the environment is rendered less favorable for the pest species.

Cultural controls were very likely the earliest control measures developed and applied by man. They include *sanitation* or cleanup of crop residues which may serve as sources of infestation, changes in planting schedules, either through *crop rotation* or use of *mixed crops* grown in relatively small plots rather than solid stands, changes in *planting or harvest time, strip cropping* or harvesting of only parts of a field at one time, use of *tillage* or working of the soil, *clean culture* or elimination of volunteer hosts, weeds, litter, or cover crops, and the use of *trap crops* which attract the pest away from the production crop. There are many variations on these basic practices, some of the more effective of which are described briefly below.

Sanitation

Sanitation has included the plowing under of infested plants after harvest, as has been done with cotton for control of the pink bollworm, the destruction of prunings of horticultural crops, citrus, apple, peach, etc., to destroy twig and branch inhabiting pests, the gathering and removal of fruit drops or mummies (stick-tights) which harbor overwintering infestations. In forests, the disposal of slash resulting from logging activity reduces infestations of bark beetles.

Crop rotation

Since ancient times crops have been rotated, that is planted in a sequence of species, for purposes of rejuvenating soil fertility (e.g., through use of legumes), suppressing soil erosion, and reducing pest infestations. Crop rotation, which is practicable mainly with field crops, has been used successfully as a means for suppressing plant-specific, soil-borne plant pathogens such as nematodes and fungi. For example, the alternation of potato with alfalfa serves to reduce infestations of wireworms; rotation of oats with corn suppresses corn rootworm.

Mixed cropping

In ecological terms, increasing the plant diversity of the agricultural ecosystem aids in the maintenance of pest populations at moderate to low numerical levels. The principal mechanism here is the perpetuation of natural controls, such as predators and parasites, which act as mortality agents for many plant pests.

Mixed planting, in modern American agriculture, commonly takes the form of interplantings, or in certain instances the planting in adjacent fields of high and low risk crops vis-à-vis a common insect pest. For example, in California some growers practice interplanting of alfalfa strips between strips of cotton so as to attract *Lygus hesperus* out of the cotton where it can do serious damage into the alfalfa where it does not. For better natural control of certain grape pests, spider mites for example, interplantings of Johnson grass or sudan grass between rows of vines are sometimes provided (Flaherty 1969, Flaherty et al. 1971).

Timing of planting or harvest

Early planting of corn in northerly regions reduces chances for infestation of *Heliothis* spp. and fall armyworm, which can only overwinter in the southern United States and thus must move northwards each summer. Early corn planting in the southern states reduces infestations of the southwestern cornstalk borer. The late planting of winter wheat each fall markedly reduces oviposition by Hessian fly.

Early harvest and subsequent plant destruction in cotton fields for control of pink bollworm and cotton boll weevil has been mentioned above. Early ripening walnuts are much less susceptible to codling moth in California. Early cutting of the first two crops of alfalfa in northern California destroys a substantial portion of the alfalfa weevil, *Hypera postica,* infestations, while the pest is still in the larval or pupal stage and cannot tolerate the loss of food or the physical shock of excessive dryness and heat resulting from the cutting activity.

Strip harvesting

Of value only with field crops such as alfalfa which provide several cuttings of hay in a season, strip harvesting, as contrasted to solid cutting of entire fields, allows the perpetuation of natural-enemy populations and consequently their maintenance of pest populations at relatively low levels (Stern et al. 1964). (See Figure 21.) However, when alfalfa fields are cut in their entirety at one time, more economical use is made of harvest crews and equipment, and irrigation is facilitated. But under the "solid cut" practice there is an enormous destruction of the resident insect fauna, including the complex of natural enemies. And, as is so often the case, as regrowth occurs the first insects to reoccupy the field in numbers are the plant pests. Only later do the predatory and parasitic species return and increase in numbers, often too late. By cropping the fields in alternative cycles, the insect fauna is maintained in the field and the new alfalfa growth is thus simultaneously populated by both pests and natural enemies. Pests which have been maintained under good

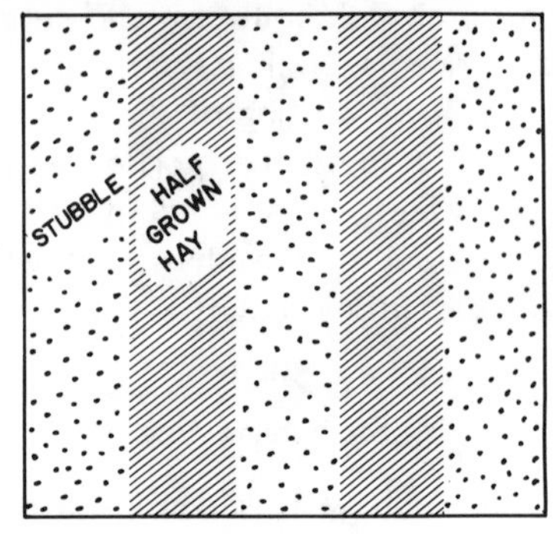

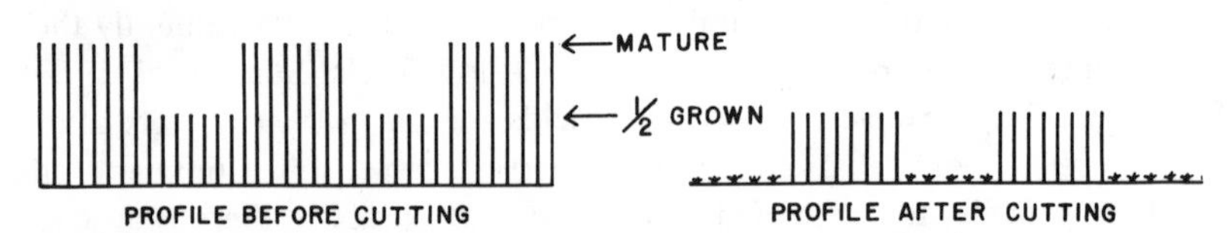

Figure 21. California alfalfa field which is undergoing the strip harvest method of cropping. Photograph and drawing courtesy of V. E. Stern, University of California, Riverside.

control by this practice are the pea aphid, *Acrythosiphon pisum*, the spotted alfalfa aphid, and several lepidopterous species, including the alfalfa caterpillar, *Colias eurytheme* Boisduval, the western yellow-striped armyworm,

Spodoptera praefica (Grote), and the beet armyworm, *Spodoptera exigua* (Hübner).

Tillage

The cultivation of the soil in and around crop plants serves to destroy numerous soil pests, either through mechanical injury or exposure. Cultivation is used principally to destroy weeds (pest plants) and to aerate and maintain proper water-penetrating properties of the soil. But the practice in many instances also destroys certain major insect pests. Fall plowing destroys overwintering European corn borers, *Ostrinia nubilalis* (Hübner), larvae of the wheat-stem sawfly, *Cephus cinctus,* and cocoons of the grape berry moth, *Paralobesia viteana* (Clemens).

Clean culture

The destruction of all weeds and border plants near a crop field and along farm roads and irrigation ditch banks and the like is a common practice in many parts of the United States. The benefits of such a practice are debatable.

On the one hand, such "scorched-earth" practice eliminates many sources of pest infestations, or pest overwintering sites, and has been claimed to reduce populations of the sorghum midge, *Contarinia sorghicola* (Coquillett), the squash bug, *Anasa tristis* (DeGeer), the harlequin cabbage bug, *Murgantia histrionica* (Hahn), the green peach aphid, *Myzus persicae* (Sulzer), and the Japanese beetle, *Popillia japonica.* The elimination of these weeds and "escaped" crop plants prevents the pest from maintaining itself at low levels at the periphery of the field or orchard, from which reservoir the new growth of the agricultural crop is invaded as the growing season advances.

On the other hand, the elimination of all weeds and other plants from the periphery of the agricultural field destroys habitats, alternate hosts, food sources, and overwintering refuges of a great variety of natural enemies and other beneficial insects (pollinators, etc.). Predators and parasites can be harbored on colonies of their preferred prey or hosts, ready to move with them into the fields each spring. Where these natural enemies have important roles in the biological control of the given pest, the interference of their stability and presence by clean-culture practices leaves the field to the exclusive play of the pest species. The consequences have been well documented: rapid pest population buildup following reinvasion, with the natural-enemy populations following much too late to prevent a great deal of early and midseason pest damage.

Because there are both positive and negative values following the practice of clean culture, each pest situation must be worked out on its own merits. Certain phases of current research in biological control attempt to retain and augment natural-enemy populations in agricultural situations, particularly with reference to field crops, by maintenance of overwintering sites, alternative host habitats, and predator "nesting sites," either by fostering weedy or bramble-type hedgerows on roadsides, or by provision of artificial refuges, bird boxes, wasp nests, coccinellid hibernation devices, and the like.

Trap crops

Trap crops are plantings in or adjacent to the agricultural field on which the pest population congregates. These trap crops are, for this reason, composed of highly preferred host plants. Since the trap crop is primarily of use as a pest control means, it may be the site of heavy pesticide treatment, or destruction by fire or the plow, or any other procedure aimed at pest elimination.

In Hawaii, trap crops of corn surrounding fields of melons or squash prove highly attractive to adult melon flies. Treatment of the corn rows with pesticides kills many pests, yet leaves no chemical residues on the crop, nor does it interfere with the natural-enemy fauna present in the field.

As has already been mentioned, the use of alfalfa strips within cotton fields serves to attract and concentrate lygus bugs which would otherwise attack and injure the cotton plants.

THE STERILE-INSECT CONTROL METHOD

The sterile-insect control method, also called the autocidal or self-destruction technique, is one of the most ingenious pest control procedures yet developed (Anon. 1969, LaChance et al. 1967). In brief, it involves, first, artificially sterilizing large numbers of pest individuals by means of gamma radiation or chemosterilants, and then releasing them into an untreated or wild population habitat, where the treated adults mate with wild adults. The result is a substantially lowered progeny production for the next generation. By repeating this procedure for a series of consecutive generations the wild population declines to eventual annihilation.

For the technique to be applied, as it was first successfully done against the screwworm fly, *Cochliomyia hominivorax* (Coquerel) in the southeastern United States in 1958, knowledge must be developed concerning (1) sterilizing the adults without at the same time reducing their competitive reproductive vigor relative to untreated adults, (2) rearing or otherwise accumulating very large numbers of the species for treatment and release, (3) accurate censusing of the population of the target insect, with respect to both absolute numbers and seasonal numerical fluctuations, (4) estimating the rate of potential population increase from generation to generation in order to arrive at suitable "overflood" ratios of treated to wild adults, (5) ascertaining that any potential harm which might result from the release of large numbers of pests will not outweigh the benefits to be realized by any subsequent annihilation or long-term suppression of the species, and (6) determining that the costs of carrying out such a control program relative to the costs of other pest suppression techniques, or to the value of the protected crop, remain favorable.

The possibility that the sterile-insect technique can literally eradicate an entire pest population from an area, and thus provide a permanent solution to the particular pest problem under consideration, has attracted a great deal of interest, theoretical attention, and practical evaluation of the several aspects of the technique. However, at the present time, 1972, the sterile-insect suppression technique has been applied in full scope in only a few situations. The limitations are of two kinds—technical and ecological.

The sterile-insect technique of pest control has been applied successfully against the screwworm fly on the island of Curacao, in Florida, and in the southwestern United States. In the first two localities the programs led to complete annihilation, in the latter area to a continuing program of pest suppression to subeconomic levels. On the small island of Rota, near Guam, in the western Pacific, the sterile-insect technique has led to the eradication (at least for a number of years) of two different species of fruit flies, the melon fly, *Dacus curcubitae* Coquillett, and the Oriental fruit fly, *D. dorsalis.* While tropical fruit flies such as these are noteworthy for their very high abundance in areas normally infested, in these two cases when eradication was undertaken on Rota the two species were present at unusually low levels, because of scarcity of host fruits and, in the case involving the melon fly, because of prior suppression with a toxic bait.

The sterile-insect procedure has also been used in pest prevention. In this modification, sterilized insects are released in an area not actually infested, but yet subject to potential, or even occasional, invasion by that particular species. Such has been the application against the Mexican fruit fly, *Anastrepha ludens* (Loew), a pest which occurs normally in many places in Mexico, but which occasionally reaches and sometimes even crosses the international border at California. Continuous releases of sterilized adults in a zone along the border serves to overwhelm any movement of the wild fly population northward. The same approach is also being used in the southern San Joaquin Valley of California as a preventive measure against spread of the very destructive cotton pest, the pink bollworm, *Pectinophora gossypiella.*

The sterile-insect technique, using either irradiated or chemosterilized individuals, has been applied on certain occasions, mostly on an experimental basis, against small segments of large pest populations with usually positive results in terms of reduction of numbers. However, such results by themselves can be misleading relative to potential annihilation or economic control. Unless the treated subpopulation is isolated (or isolatable) from other subpopulations, or unless the technical feasibility for annihilation or control of the entire population has been established, such results do not themselves constitute the successful application of the technique.

Drawbacks to the procedure

Effective sterilization of the adult of a species must be attained without impairing suitable levels of reproductive competitiveness in the insect. In many cases, preliminary research to establish such post-sterilizing reproductive competitiveness indicates that the method is unfeasible.

The ability to rear the pest organism by artificial means in very large numbers constitutes another technical barrier. It so happens that in the successful autocidal control programs the species concerned lent themselves to mass culture. For the screwworm eradication program in Florida, production levels of over 50 million flies per week were accomplished. For the southwestern United States-Mexico suppression program, 100 to 150 million flies per week were reared. For eradication of the two *Dacus* species from Rota, sterile releases amounted to about 10 million per week. In all these cases, the pests were amenable to rearing in mass situations, on artificial media, under conditions of excessive crowding.

Hence the sterile-insect technique will very likely be impracticable for pests which cannot be sterilized without excessive behavioral alteration, or for those species which can only be reared on living plant material, those which are highly cannibalistic as larvae, or those of excessive susceptibility to disease which must be reared under aseptic conditions.

Nevertheless, there are a number of species, besides the ones mentioned above, which may very well be susceptible to the sterile-insect technique. Given the appropriate ecological circumstances of isolation and wild population accessibility, the Mediterranean fruit fly, *Ceratitis capitata,* the codling moth, *Laspeyresia pomonella,* the boll weevil, *Anthonomus grandis* Boheman, and similar major pest species may well be appropriate for such a control approach. But even these programs may not be feasible because of cost. Some of the programs will be extremely costly (e.g., it is estimated that complete eradication of the boll weevil, if accomplished, will cost $300 million). Is society willing to pay this staggering price, especially if alternative methods such as cultural or biological control, the use of resistant varieties, or integrated control can accomplish satisfactory results much more cheaply?

GENETIC CONTROL OF PESTS

The genetic control of pests involves the alteration of the genetic makeup of pest organisms such that they become less vigorous, less fecund, or genetically sterile as a consequence of hybridization, resulting in their decline in numbers to either low densities or extinction (Anon. 1969). The term *genetic control* has been amplified by some to include host plant resistance, on the one hand, or the sterile-insect principle, on the other. We prefer to restrict the scope of genetic control to some artificial influence by man on the gene composition of individuals. Host plant resistance and sterile-insect methods are treated in separate sections above.

Many basic properties of insects lend themselves to genetic manipulation. Species of insects appear in numerous polytypic forms, biotypes, races, and strains. Populations of insects are easily altered by selection, either natural or artificial, this genetic plasticity being evidenced by their highly varied and variable adaptations to different ecological situations. Coupled with this evidence of their genetic plasticity is the suitability of their life histories for genetic manipulation. They have short generation times and relatively high reproductive potentials, which make them excellent subjects for selective breeding and genetic experimentation.

Research on the genetic control of insect pests has centered on the following: (1) hybrid sterility, (2) cytoplasmic incompatibility, (3) conditioned lethals, and (4) growth alterations. None of these has yet led to field testing, let alone actual pest control. However, research on several lines is presently under way, and practical applications may well turn up within the next decade or so.

Hybrid sterility

Hybrid sterility is the result of the hybridization of two different, related species, with the production of sterile progeny. Where these hybrids are fully viable, they remain in the environment to compete with normal individuals, with a consequent reduction in the breeding population.

This scheme requires the determination of species which will cross, one being the target pest and one a nonpest. Mass culture and release of the nonpest species in the target pest area will bring the control method into action.

This method has been proposed for the two tsetse fly species, *Glossina swynnertoni* Austen and *G. morsitans* Westwood. Where, as in this case, both species are pests, mass culture and hybridization will have to be carried out in "mosquito factories" with release only of the sterile hybrids. To this extent, the procedure is equivalent to the sterile-insect control method.

Cytoplasmic incompatibility

Certain strains of the same species when crossed produce either sterile offspring, or fertile and sterile offspring together (resulting from reciprocal crosses), or sterile males. The sterility in any case is the result of the destruction of the sperm as it penetrates the ovum.

This technique has been proposed for a number of mosquito species where these incompatible strains have been found, including *Culex pipiens pipiens* Linnaeus and *Aedes scutellaris* (Walker). One of two modifications can be used: production and release of males of one type into the area of occupancy of an incompatible type, or production and release of the hybrid, which will compete sexually with the target pest.

Conditional lethals

Certain alleles of genes controlling adaptiveness to temperature, for example, produce normal effects under certain temperature ranges, but at higher or lower temperatures produce nonadapted or lethal effects. Again, this scheme derives from observations of mosquitoes, in this case the yellow fever mosquito *Aedes aegypti* (Linnaeus). A strain of this species has been found which when reared at 27-28°C produces normal adults in a 1:1 sex ratio, while when reared at 30-34°C half the progeny were normal females and the rest intersexes. These latter resemble normal females, and will mate and receive sperm, but will produce no eggs. Any control procedure using this technique will require the introduction of large numbers of the maladapted strain into the target pest environment.

Whitten (1970) has proposed a method for using conditionally lethal genes of the sort described above. In his procedure, the strain of pest possessing the conditional lethal allele is irradiated to produce chromosome translocations. So long as this translocation strain is maintained in a homozygous state, it remains reproductively fertile. Mass culture and release of such a strain enables it to hybridize with the wild or pest strain to produce the heterozygous state. These hybrids are sterile. By suitable overflooding of the wild population it eventually is totally displaced. Then, with a shift in the seasonal temperature, the conditional lethal comes into play and results in the self-destruction of the released strain. The overall result could be the full elimination of both strains from the target environment.

Growth alterations

In some species of insects, the presence or absence of diapause is known to be under genetic control, as for example the silkworm, *Bombyx mori* (Linnaeus), and the cotton boll weevil, *Anthonomus grandis.* By use of non-diapause genes in a "translocation" strain, a normal, diapausing strain can be displaced, as per Whitten (1970), discussed above. When the winter (or summer) diapause season arrives, the released strain then is annihilated.

Another procedure for genetic manipulation of a diapausing population has been considered in the case of the field cricket, *Teleogryllus commodus* (Walker) in Australia. In the southerly, more temperature regions this species produces diapausing eggs; in the more tropical north it produces only non-diapausing eggs. Northern males, crossed with southern females, produce nondiapausing eggs. Hence, mass culture and release of northern males into southern infestations should result in the production of numerous eggs unable to carry over the winter. Repeated releases as in the sterile-male technique conceivably could produce substantial and even complete reduction of the southern strain of the field cricket.

REVIEW AND RESEARCH QUESTIONS

1. Give four examples of pests which have been controlled by means of host plant resistance. What are the three mechanisms by which plants resist insect attack?

2. Discuss the various methods of cultural control used to suppress insect pests.

3. What are the requirements for use of the sterile-insect control method?

4. How can genetic aberrations be used to control pests?

BIBLIOGRAPHY

Literature cited

Anonymous. 1969. *Insect-pest management and control.* Nat. Acad. Sci. Publ. 1695 (v. 3 in series Princ. Plant and Anim. Pest Contr.). Washington, D.C. 508 pp.

Beck, S. D. 1965. Resistance of plants to insects. *Ann. Rev. Ent.* 10: 207-232.

Flaherty, D. 1969. Ecosystem trophic complexity and Willamette mite, *Eotetranychus willamettei* Ewing (Acarina: Tetranychidae), densities. *Ecology* 50(5): 911-916.

Flaherty, D. L., C. D. Lynn, F. L. Jensen, and M. A. Hoy. 1971. The influence of environment and cultural practices on spider mite abundance in southern San Joaquin Valley Thompson seedless vineyards. *Calif. Agric.* 25(11): 7-8.

LaChance, L. E., C. H. Schmidt, and R. C. Bushland. 1967. Radiation induced sterilization. In *Pest control, biological, physical, and selected chemical methods,* ed. W. W. Kilgore and R. L. Doutt, Chap. 4. New York: Academic Press, 477 pp.

Painter, R. H. 1951. *Insect resistance in crop plants.* New York: Macmillan, 520 pp.

Painter, R. H. 1960. Breeding plants for resistance to insect pests. In *Biological and chemical control of plants and animal pests.* Amer. Assoc. Advan. Sci., Washington, D.C.

Stern, V. M., R. van den Bosch, and T. F. Leigh. 1964. Strip cutting alfalfa for lygus bug control. *Calif. Agric.* 18: 406.

Whitten, M. 1970. Genetics of pests in their management. In *Concepts of pest management,* ed. R. L. Rabb and F. E. Guthrie, pp. 119-137. North Carolina State Univ., Raleigh, N.C., 242 pp.

10 The Future of Biological Control

Biological control, as measured by the permanent suppression of pest species, ranks as one of the most effective insect and weed control tactics. No one really knows how much benefit has been derived from pest control effected by natural enemies. DeBach (1964) estimated that between 1923 and 1959 classical biological control programs costing about $4.3 million benefitted the California agroeconomy alone in the amount of about $115 million. Unquestionably, many more millions of dollars in direct benefits have accrued in the ensuing years. If we then add the benefits from programs carried out in California before 1923 and those from the many successful programs conducted elsewhere in the world, it is apparent that savings amounting to several hundreds of millions of dollars have been realized from classical biological control. And then, too, we must consider the contribution of biological control to environmental quality.

But natural-enemy introduction programs are only partially responsible for the benefit realized from biological control; naturally occurring parasites, predators, and pathogens add immensely to the total. It is impossible to even guess at the financial benefit realized from naturally occurring biological control, but as was mentioned in Chapter 1, it is not inconceivable that this great natural force may well be crucial to our very survival.

What then, of the future utilization and manipulation of natural enemies? Has the cream been skimmed or can we expect even greater benefits from biological control? We take the optimistic view. Indeed, it is our opinion that if expanded support and effort are forthcoming, even greater rewards can be expected, especially since this would permit unprecedented exploitation of naturally occurring biological control as well as greater efficiency in natural-enemy importation. The following discussion is a brief outline of our thoughts on the future trends and developments in biological control.

CLASSICAL BIOLOGICAL CONTROL

Natural-enemy importation, which has been going on for more than eighty years, has hardly approached its full potential. Even in the United States, where for decades there has been a relatively high level of activity, such major pests as the codling moth, boll weevil, and green peach aphid have never been serious subjects of natural-enemy introduction efforts. In addition, there are literally scores of exotic pests of lesser importance against which there has been little effort in the area of natural-enemy introduction. For example, of the scores of pest aphid species in this country, almost all of which are exotic, only a handful have been targets of classical biological control.

There are many countries in the world where there has been virtually no effort in biological control. In some countries this neglect has been largely due to skepticism on the part of policy-making entomologists, while in others it has resulted from inadequate resources, both financial and technical. Some governments have met this latter problem through membership in the Commonwealth Institute of Biological Control, the International Organization of Biological Control, or in programs sponsored by the Food and Agricultural Organization (FAO) of the United Nations. But this participation requires financial commitment and such funds are not always available.

Even in those countries with the most active classical biological control programs, support has been far from lavish. For example, in the United States, federal expenditures for classical biological control against the whole spectrum of exotic pests probably amounts to little more than half a million dollars annually. This is to be contrasted with the massive amounts of federal money expended on eradication or area control programs against such pests as the fire ant ($10 million in 1971 alone), gypsy moth, or white fringed beetle, or the hundreds of millions of dollars which society expends yearly on chemical control of insects and the monitoring programs that the chemicals require.

But biological control's slim fiscal fare may well be in line for augmentation as increasing public concern over pesticides generates greater pressures for alternative controls. And if increased support is forthcoming the numbers of biological control successes will certainly increase. As DeBach (1964) has pointed out, the degree of success in classical biological control is directly proportional to the amount of effort given to natural-enemy introduction. In other words, those countries or agencies which have made the greatest effort have reaped the most benefit. Thus, among the United States, Hawaii and California have had by far the greatest success in classical biological control largely because they have pursued this tactic much more vigorously than the other states. It would seem, then, that if DeBach's formula were applied on a global basis, and support for natural-enemy introduction was doubled, say, there might well be a doubling of the number of successes. In fact, with our increased understanding of entomophagous insects and improved techniques in the selective procurement, production, and colonization of natural enemies, the success ratio might be substantially higher than what it has been historically. But whatever the relative degree of success, there is abundant opportunity to add significantly to the triumphs of classical biological control through increased effort. Society, then, must decide whether it wants to invest in such an expanded effort with its real potential for substantial economic and ecological benefit. In light of past success it is difficult to envisage anything but increased public support for classical biological control in this era of ecological concern.

NATURALLY OCCURRING BIOLOGICAL CONTROL

There can be little doubt that research on naturally occurring biological control will increase in the future. The proliferating integrated control programs with their heavy reliance on natural enemies is a clear indication of things to come (Huffaker 1971). And as time passes and information and experience are gained, even better use will be made of naturally occurring biological control. As this occurs, a very important windfall will result, namely increased knowledge of entomophagous arthropods. Thus, we can expect to gain continually expanding insight into the systematics, biology, ecology, physiology, ethology, host ranges, and nutrition of parasitic and predaceous arthropods, and an improved understanding of host-natural enemy interrelationships. This information, in turn, will have a positive feedback effect in that it will increase even further our efficiency in the use of naturally occurring biological control.

SPECIAL MANIPULATIONS OF NATURAL ENEMIES

Special manipulation of natural enemies can perhaps be termed the third dimension of biological control. Natural-enemy manipulation is one of the most neglected areas of biological control, and yet one of the most promising. Improved mass propagation techniques, use of artificial diets, and better understanding of natural-enemy ecology are the basis for this optimism. Practices of particular promise are periodic colonization of natural enemies, their nutritional augmentation, and the development of artificial or cultural manipulations designed to retain and augment their populations in the crop environments.

Innovation in production techniques, utilization of artificial nutrients or other habitat improvements, precise timing of releases, and the use of more effective natural-enemy species or strains should all contribute importantly to improved efficacy in the periodic colonization tactic. Studies in England and Europe (Hussey and Bravenboer 1971) have shown that there is an excellent potential for this technique in glasshouse pest control. Pest problems in a variety of high value crops (e.g., strawberry, vegetables, ornamentals), and in parks and home gardens also seem particularly amenable to the periodic colonization technique. However, the economics and logistics of natural-enemy mass propagation and colonization present a formidable obstacle to the extensive adoption of this technique. On the other hand, since this is an aspect of biological control that would involve the marketing of a product (natural enemies), it might well generate the kind of capitalization and production know-how that will permit profitable large-scale propagation of natural enemies. In fact, certain traditional pesticide-producing companies in Japan and Europe have already entered the natural-enemy production field on a limited scale.

Advances in the nutrition of entomophagous insects are being made along two lines. The first of these is the development of synthetic or semisynthetic diets as substitutes for living hosts in the mass propagation of natural enemies. This use of synthetic nutrients may well be a key factor in the practical utilization of insectary-grown natural enemies as "biotic insecticides." Provision of foodstuffs for field populations of natural enemies is the second promising avenue in the nutrition area. Here, cheap but complete foodstuffs are sprayed on vegetation or otherwise presented to the natural enemies either to attract them to desired places, sustain them during periods of natural food scarcity, or to increase their reproduction rates (Hagen et al. 1971). In some cases all three features are met by distribution in the field of a given nutrient. Our colleague K. S. Hagen has been a pioneer in this field, and already a dairy by-product, developed by him as a predator food supplement, is being used on a commercial basis in California.

Environmental manipulations to preserve and augment natural enemies have been little exploited. This perhaps derives from the prevailing ignorance of the true significance of naturally occurring biological control. But a drastic change is clearly in the offing. Studies on strip harvesting of alfalfa in California leave little doubt that universal adoption of this practice in the state would strikingly reduce pest insect problems in this $200 million crop. The provision of nesting places, diapause quarters, nectar plants, and sheltering sites and other tactics have all been effective enhancers of natural-enemy activity in a spectrum of crops (van den Bosch and Telford 1964). This promising record is a clear challenge to expand our efforts in this area of natural enemy augmentation.

PEST MANAGEMENT AND BIOLOGICAL CONTROL

In recent years a broadened concept of pest control has emerged, termed *pest management.* It is derived from the ecologically based integrated control approach to pest suppression (see Chapter 8), but is a much expanded concept which includes the multidisciplinary consideration of all types of pests, insects, plant pathogens, nematodes, and weeds whenever these are of concern in the same crop environment, as well as utilization of knowledge about the economics of crop production.

Because of the holistic framework, and because decisions made by or for the grower about control of insect pests, for example, are influenced by or will effect decisions made about plant disease or weed control, pest management programs will of necessity often rely on a systems analysis approach for arriving at optimal programs and decisions.

Biological control will certainly be a major component of pest management programs. This will be so not only because biological control is a commonly used component of integrated control, but also because the concept and philosophy of biological control conforms closely with the concept and philosophy of pest management. Consequently, as the pest management idea becomes implemented through research and pilot development, we can expect to see an expanded need for and interest in biological control work as described herein.

CONCLUSION

Biological control, a natural phenomenon, is a great biotic force which helps regulate insect populations and those of myriad other organisms as well. Like so many of our natural resources, biological control can either be squandered, with detriment to us and our environment, or it can be conserved, augmented, and manipulated with beneficial effect.

We have it within our power to choose between these alternatives. The nature of that choice may have important bearing on our future success as a competing species on this planet.

REVIEW AND RESEARCH QUESTIONS

1. Classical biological control has been going on for more than eighty years. Does this mean that further work would be unlikely to bring any further successes?

2. What would be needed to apply biological control in developing countries?

3. Of what value are studies on the presence and effectiveness of naturally occurring biological control?

4. How can we manipulate natural enemies to increase their value?

5. What is pest management?

BIBLIOGRAPHY

Literature cited

DeBach, P. 1964. Successes, trends and future possibilities, Ch. 24, pp. 673-713 in *Biological control of insect pests and weeds.* P. DeBach (ed.) London: Chapman and Hall.

Hagen, K. S., E. F. Sawall and R. L. Tassan. 1971. The use of food sprays to increase effectiveness of entomophagous insects. In *Proc. Tall Timbers Conf. on Ecol. Anim. Control* No. 2, Tall Timbers Res. Inst., Tallahassee, Fla. 1970. Pp. 59-81.

Huffaker, C. B. (ed.) *Biological control.* New York-London: Plenum Press. 1971.

Hussey, N. W. and L. Bravenboer. 1971. Control of pests in glasshouse culture by the introduction of natural enemies. Chap. 8, pp. 195-216 in *Biological control.* C. B. Huffaker (ed.) New York: Plenum Press.

van den Bosch, R. and A. D. Telford. 1964. Environmental modification and biological control. Chap. 16. 16, pp. 459-488 in *Biological control of insect pests and weeds.* P. DeBach (ed.). London: Chapman and Hall.

Glossary

adelphoparasitism • self-parasitism, that form of parasitism where one sex develops parasitically in the body of the opposite sex

agroecosystem • the ecosystem composed of cultivated land, the plants contained or grown thereon, and the animals associated with these plants

alien species • an organism which has invaded and is growing in a new region

arrhenotoky • that pattern or mode of reproduction where progeny of both sexes are produced by mated females, the egg when unfertilized producing a viable, haploid male and when fertilized producing a viable, diploid female

biological control • the regulation of plant and animal numbers by biotic mortality agents (natural enemies); also, the use by man of natural enemies to control (reduce) the numbers of a pest animal or weed

biotic agents • living environmental factors which favor the well-being or bring about the premature death of animals and plants

broad-spectrum pesticide • a material of such broad toxicity that it kills not only a range of pest species but also many nontarget species, such as natural enemies, honeybees, birds, and other forms of wild life

carnivore • an animal which feeds on other live animals

cleptoparasitism • a case of multiple parasitism where a parasitoid preferentially attacks a host already parasitized by another species rather than an unparasitized host

colonization • the controlled release of a quantity of biological control agents in a favorable environment for the purpose of permanent or temporary establishment

community • a characteristic assemblage of interacting populations of species in a particular habitat

competitition • the interference which occurs among individuals utilizing a common resource which is insufficient in quantity to satisfy the needs of all consumer • a heterotrophic organism or population, usually animal, fungus, or virus, which utilizes dead or living organic matter as food

cultural control • a pest control method where normal agronomic practices—tilling, planting, crop spacing, irrigating, harvesting, waste disposal, crop rotation—are altered so that the environment is less favorable or unfavorable for the pest

decomposer • a heterotrophic organism which utilizes dead organic matter as food, decomposing it into more simple substances

delayed density-dependent mortality • mortality inflicted on the members of a population the magnitude of which is determined by the density of the population at some time in the past

density-dependent mortality • mortality inflicted on the members of a population, the degree of which is related to or affected by the density of the population

density-independent mortality • mortality inflicted on the members of a population, the degree of which is unrelated to or unaffected by the density of the population

deuterotoky • that pattern or mode of parthenogenetic reproduction where progeny of both sexes are produced by unmated females

direct density-dependent mortality • mortality inflicted on the members of a population the magnitude of which increases as the current density of the population increases and which decreases promptly as the current density of the population declines

direct hyperparasite • a parasitoid which searches out and deposits its egg in or on the body of its parasitic host, which may or may not be contained in the body of its own living host

economic threshold • the density of a pest population below which it fails to cause enough injury to the crop to justify the cost of control efforts

ectoparasite • a parasitoid which develops externally on the body of the host

encapsulation • an immune response of certain insect hosts to nonadapted parasitoid eggs and young larvae wherein the immature parasitoids become covered with and eventually encysted by certain host blood cells, with death of the parasitoids soon following

endoparasite • a parasitoid that develops within its host's body

entomophagous insect • a predatory or parasitic insect which feeds on other insects

entomophagy • the consumption of insects by other animals

epizootic • the widespread or simultaneous occurrence of a disease in a large proportion of an animal population

establishment of natural enemy • the permanent occurrence of an imported natural enemy in a new environment after the act of colonization has been carried out

exotic species • an organism which evolved in one part of the world and which now occurs either accidentally or intentionally (by man) in a new region

factitious host • an easily grown plant or animal species used as a host for the mass culture of a natural enemy in the insectary but which is not attacked by this enemy in nature

fecundity • the number of eggs or offspring the females of a species can produce during their lifetime

food chain • a trophic path or succession of populations through which energy flows in an ecosystem as a result of consumer-consumed relationships

food web • a complex of branching, joining, or diverging food chains which connect together the various populations in an ecosystem

genetic control • a pest control method which makes use of selected strains of the target species possessing chromosome aberrations or other genetic abnormalities such that, when released into the target population, mating with wild (normal) individuals results in production of sterile or less viable progeny

gregarious parasite • a parasitoid whose food requirements are such that from several to many can develop simultaneously in or upon the body of the host

habitat • a physical portion of the environment within which a population is dispersed

haploid parthenogenesis • the situation where the unfertilized egg hatches and develops normally to produce a viable, male adult whose cells contain only the haploid number of chromosomes

herbivore • an animal which feeds on live green plants

host plant resistance • a method of pest control in which crop varieties are used which are resistant, tolerant, or unattractive to the pest

host specificity • the degree of restriction of the number of different plant or animal species which can serve as a food source for herbivorous or carnivorous species

hyperparasite • an insect which is parasitic in or on another parasitic insect indirect hyperparasite • a secondary parasitoid which searches out and deposits its egg in the body of an unparasitized, nonparasitic host, the egg usually remaining undeveloped until the nonparasitic host is subsequently parasitized by a primary parasitoid which then serves as host for the secondary

innoculative releases • the repeated colonization of relatively small numbers of a natural enemy for purposes of building up a population over several generations

integrated control • an ecologically based, pest population management system which uses all suitable techniques to reduce or so manipulate the pest population that it is prevented from causing economically unacceptable injury to the crop

inundative release • a colonization of large numbers of a natural enemy for the immediate purpose of inflicting prompt mortality on the pest population

inverse density-dependent mortality • mortality inflicted on the members of a population the magnitude of which decreases as the current density of the population increases and which increases as the current density of the population declines

mass culture • the propagation in an insectary of very large numbers of a biological control agent, often on a continuous basis over a period of months or years

microbial insecticide • a material composed of or containing pathogenic microbes or their toxic products for use in controlling insect pests

monophagy • the restriction of an animal to the consumption of but one species of food organism

mortality factor • a factor or agency in the environment which causes the premature death of an organism

multiple parasitism • the situation where more than one parasitoid species occurs simultaneously in or on the body of the host

natality • the properties, reproductive or dispersive, which enable a population to increase in numbers

natural control • the collective action of environmental factors to maintain the numbers of a population within certain upper and lower limits over a period of time

natural enemy • an animal or plant which causes the premature death of another animal or plant

niche • the place or position, in both a physical and a functional sense, of a species population in an ecosystem as determined by the full complex of environmental factors impinging on and limiting the population

nonreciprocal density-dependence • mortality inflicted on a population by a biotic mortality factor whose own numbers are not changed as a consequence

oligophagy • the restriction of an animal to the consumption of a moderate range of food organisms

parasite • a small organism which lives and feeds in or on a larger host organism

parasitism • a biotic interaction involving a trophic relation between a small organism and its larger host

parasitoid • a parasitic insect which lives in or on and eventually kills a larger host insect (or other arthropod)

parthenogenesis • the production of normal progeny by unmated females

pathogen • a microorganism which lives and feeds (parasitically) on or in a larger host organism and thereby causes injury to it

pathogenesis • the causing of a state of ill health, morbidity, and in some cases the premature death, in a host organism by a pathogenic microorganism

periodic colonizations • the frequent release of small or large numbers of a natural enemy for short-term control action rather than permanent establishment

pest • a species which because of its high numbers is able to inflict substantial harm to man, domesticated animals, or cultivated crops

pest resurgence • the rapid numerical rebound of a pest population after use of a broad-spectrum pesticide, brought about usually by the destruction of natural enemies which were otherwise holding the pest in check

phytophagy • the consumption of plants or plant tissues by animals

polyphagy • the consumption by an animal of a wide variety of food organisms

population • an aggregation of similar individuals in a continuous area which contains no potential breeding barriers

predation • a biotic interaction involving a trophic association between a large or strong animal and a small or weak animal

predator • an animal which feeds upon other animals (prey) which are either smaller or weaker than itself

primary parasite • a parasitoid which develops in or on nonparasitic hosts

producer • an autotrophic organism or population, usually green plants, which procure energy from outside the ecosystem and through the process of photosynthesis convert this energy into living organic matter within the system

quarantine • the containment of an imported or immigrant individual in a special, escape-proof facility until its identity is determined and any associated, potentially hazardous organisms are eliminated

reciprocal density-dependence • mortality inflicted on a population by a biotic mortality factor whose own numbers are changed as a consequence

regulation • as related to population dynamics, the control of population density

reproductive capacity • the capacity of the individuals in a population to increase in number by the production of progeny

scavenger • animal which utilizes the dead bodies or tissues of other organisms as food

secondary parasite • a parasitoid which develops in or on a primary parasite

secondary pest outbreak • the rapid numerical increase to pest status of a scarce, noneconomic, phytophagous population after use of a broad-spectrum pesticide for control of another pest of the crop, brought about by the destruction of natural enemies which were otherwise holding the secondary pest in check

solitary parasite • a parasitoid whose food requirements are such that no more than one can develop successfully in the body of the host

stenophagy • the restriction of an animal to the consumption of but a narrow range of closely related species of food organisms

sterile-insect control • a pest control method which makes use of artificially sterilized populations of the pest to mate with and thereby interfere with normal reproductive efforts of the target species

superparasitism • the situation in which more individuals of a parasitoid species occur in a host than can survive

tertiary parasite • a parasitoid which develops in or on a secondary parasite

thelyotoky • that pattern or mode of parthenogenetic reproduction where unmated females produce only female progeny, males being unknown

trophic level • a particular step occupied by a species population in the process of energy transfer within an ecocystem

weed • an aggressive, invasive, easily dispersed plant, one which commonly grows in cultivated ground to the detriment of a crop

Supplementary Illustrations

Pea aphids mummified by *Aphidius smithi.* Photo by F. E. Skinner, University of California, Berkeley.

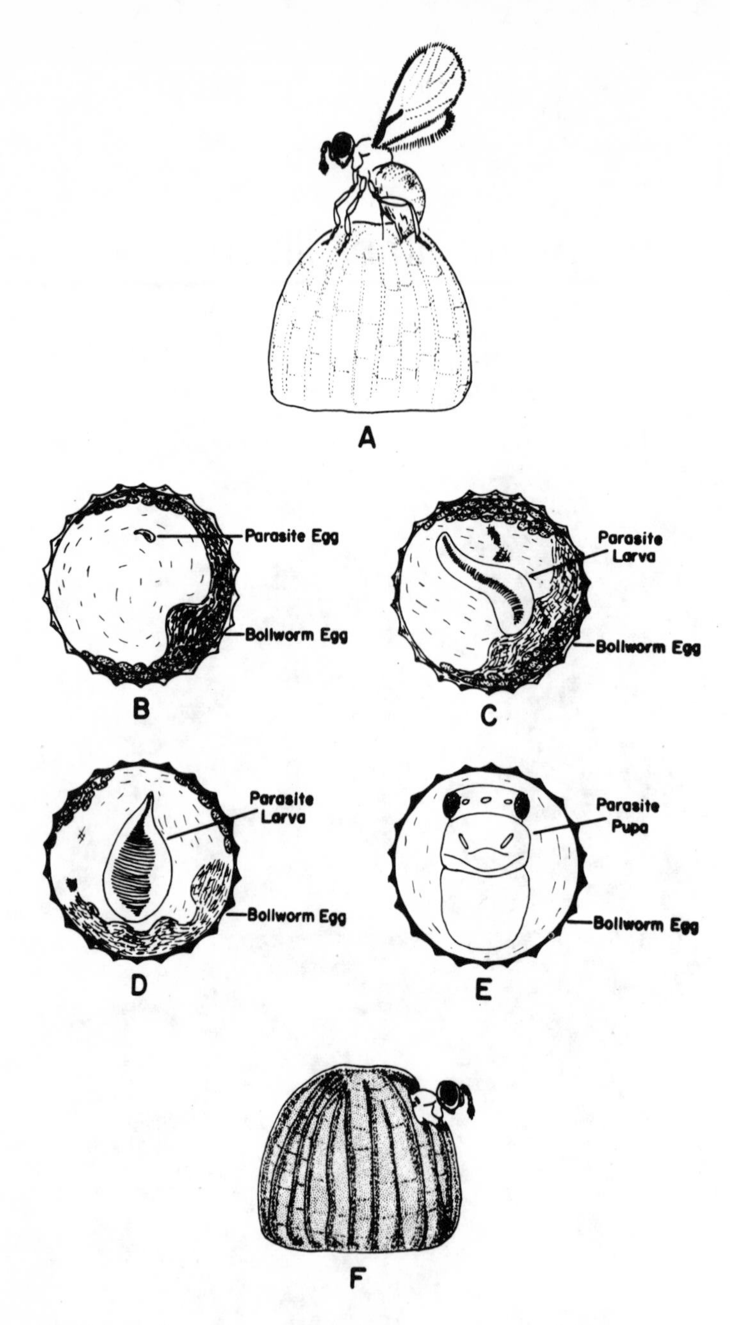

Life history of a solitary endoparasite, the egg parasite *Trichogramma* sp. (A) *Trichogramma* female ovipositing in a bollworm (*Heliothis zea*) egg, (B) *Trichogramma* egg within the bollworm egg (dorsal view), (C) and (D) larval stages, (E) pupa, (F) adult emerging through hole it has cut in the egg shell. Drawings by C. F. Lagace.

Typical predators commonly encountered in biological control work.

Hover fly (Syrphidae) larva. Photo by F. E. Skinner, University of California, Berkeley.

Hover fly (Syrphidae) adult. Photo by F. E. Skinner, University of California, Berkeley.

Typical predators (continued)

Predaceous wasp, *Polistes apachus* Saussure. Photo by Ken Middleham, University of California, Riverside.

Adult convergens lady beetle, *Hyppodamia convergens* Guerin, feeding on an aphid. Photo by Ken Middleham, University of California, Riverside.

Assassin bug, *Zelus renardii*, Kolenati, adult. Photo by Ken Middleham, University of California, Riverside.

Assassin bug, *Zelus renardii*, nymph. Photo by Ken Middleham, University of California, Riverside.

Wolf spider (Lycosidae). Photo by Ken Middleham, University of California, Riverside.

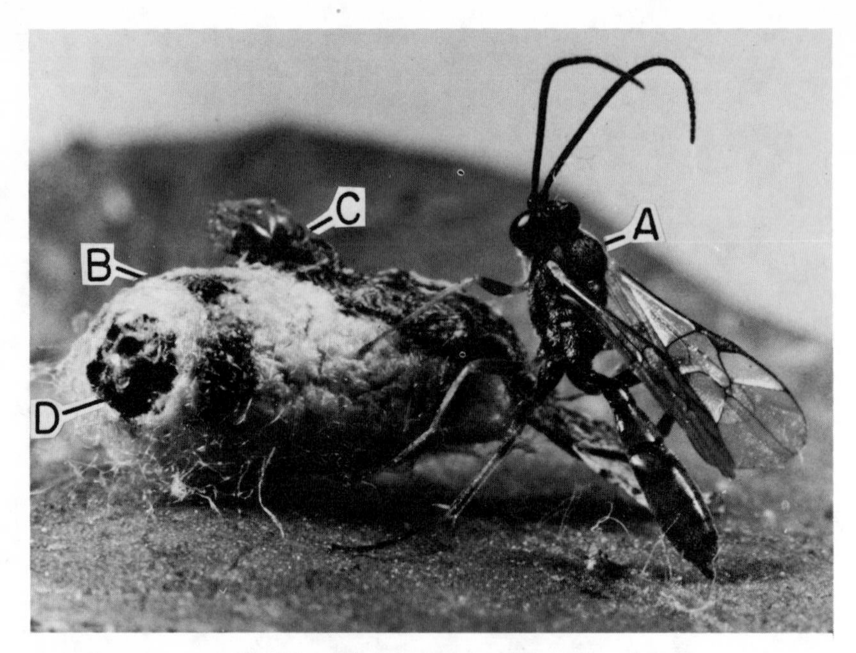

Hyposoter exiguae (Viereck), a solitary endo-parasite of lepidopterous larvae. (A) adult, (B) cocoon, (C) skin and head capsule of dead host caterpillar, (D) exit hole made by the wasp in escaping its cocoon. Photo by Ken Middleham, University of California, Riverside.

Adults of *Pentalitomastix plethoricus*, a polembryonic species (multiple embryos from a single egg) that have emerged from a single navel orangeworm. Photo by F. E. Skinner, University of California, Berkeley.

"Mummified" body of a navel orangeworm packed with hundreds of pupae of the parasite *Pentalitomastix plethoricus* Caltagirone. Photo by F. E. Skinner, University of California, Berkeley.

Leptomastix dactylopii Howard, a parasite of mealybugs. Photo by John Black & Associates.

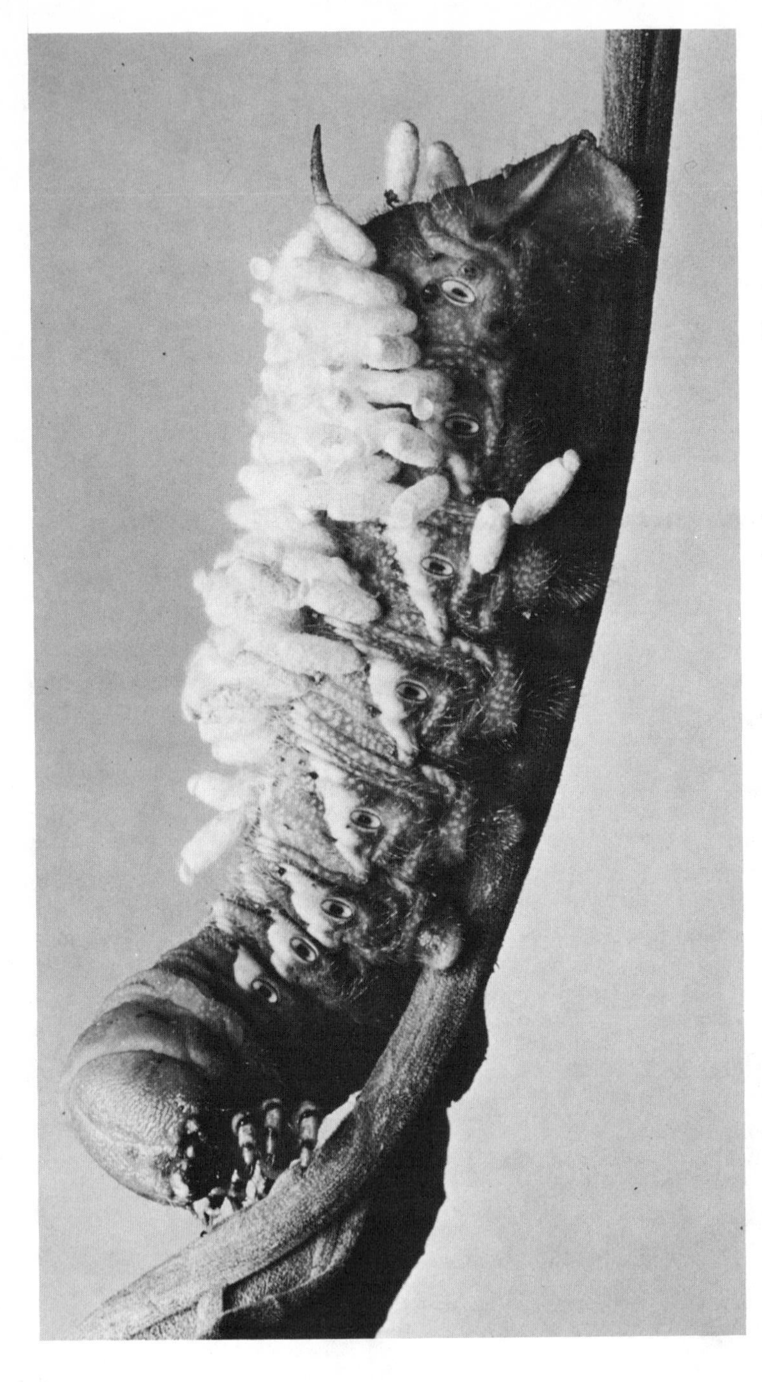

Cluster of cocoons of the gregarious endoparasite, *Apanteles congregatus* (Say) on a hornworm. The *Apanteles* larvae after reaching maturity bore through the host's integument to spin their cocoons. The hornworm, though lifelike, is mortally affected. Photo by F. E. Skinner, University of California, Berkeley.

Index